A Stereotaxic Atlas
of the Rat Brain

A Stereotaxic Atlas
of the Rat Brain

Louis J. Pellegrino
Purdue University
West Lafayette, Indiana

Ann S. Pellegrino

and

Anna J. Cushman

PLENUM PRESS • NEW YORK AND LONDON

Library of Congress Cataloging in Publication Data

Pellegrino, Louis J
  A stereotaxic atlas of the rat brain.

  Bibliography: p.
  Includes index.
  1. Brain—Atlases. 2. Rats—Anatomy—Atlases. I. Pellegrino, Ann S., joint author.
II. Cushman, Anna J., joint author. III. Title.
QL937.P45 1979   599'.3233   79-9438
ISBN 0-306-40269-6

First Printing—November 1979
Second Printing—January 1981
Third Printing—February 1986

(Second edition of A Stereotaxic Atlas of the Rat Brain)

© 1967, 1979 Plenum Press, New York

A Division of Plenum Publishing Corporation
233 Spring Street, New York, N.Y. 10013

Printed in the United States of America

A Stereotaxic Atlas
of the Rat Brain

Introduction

The first edition of this atlas was developed at the Allan Memorial Institute of McGill University in the laboratory of Dr. Robert B. Malmo. It was originally intended for use in the McGill psychology laboratories as a guide for stereotaxic placement of electrodes, cannulae, and lesions, and for assisting in the histological verification of these placements. It was published in the hope that it would be of similar use to investigators in other laboratories and in related fields. Since the publication of the first edition 12 years ago, we have come to feel that the usefulness of the atlas would be considerably enhanced by the addition of coronal sections extending caudally to include the entire brainstem and cerebellum and by the inclusion of a full set of sagittal sections. Fortunately, when the time came to consider a second edition, Dr. Joseph Altman generously made available to us the facilities of the Laboratory of Developmental Neurobiology at Purdue University, and our desire to expand the original atlas has thus been fulfilled.

MATERIALS AND METHODS

All sections were obtained from the brains of male hooded rats (280–320 g body weight) from either the Royal Victoria Hospital strain (Figs. 1–80) or the Purdue Long–Evans strain (Figs. 81–122). Anatomical differences between the hooded rat and the white laboratory rat (both are strains of the brown rat *Rattus norvegicus*) are no greater than differences between individual white rats, as far as stereotaxy is concerned. Hence, this atlas is as applicable to studies on the white rat as to work on the hooded rat (or other strains of *R. norvegicus*). Investigators who may wish to use this atlas with rats that are either larger or smaller than 280–320 g are referred to a paper by Whishaw, Cioe, Previsich, and Kolb (1977) which provides a mathematical formula for the use of this atlas with animals ranging in weight from 161 to 782 g.

The rats were sacrificed with ether or pentobarbital sodium (60 mg/kg) and perfused intracardially with 40 ml of physiological saline followed by 40 ml of 10% formol–saline. The brains used in Figs. 1–80 were refrigerated for 24 hr, then removed from the skulls and stored in 10% formalin for a minimum of 48 hr. The brains used in Figs. 81–122 were removed from the skulls immediately following perfusion and stored in 10% formalin for a minimum of 48 hr. They were then placed in a 10% formalin–20% sucrose solution for 48 hr prior to sectioning.

3

Frozen sections of the brains were obtained at a thickness of 40 μm. The brains were mounted on the object disc of the cryotome so that the coronal sections would be in the same plane as those in the de Groot (1959) atlas. Brains to be sectioned sagittally were mounted so that the plane of sectioning would be parallel to the midsagittal plane. The sections were stained with luxol fast blue for myelinated fibers and counterstained with neutral red for cell groups [modified Klüver and Barrera (1953) technique]. These sections were then photographed at 12×. The atlas consists of photomicrographs of 99 coronal sections and 23 sagittal sections and their corresponding schematic drawings. The coronal sections are at 200-μm (0.2-mm) intervals. The medial sagittal sections (0.1–3.5 mm) are spaced at 200-μm intervals and the lateral sagittal sections (3.5–5.1 mm) are spaced at 400-μm intervals.

For the first edition (Figs. 1–80), a second set of identical coronal sections was stained with thionin and used for making the schematic drawings. Since it is very difficult to obtain identical sections from two different animals, and since we have found that it is quite possible to produce drawings of good quality from luxol fast blue–neutral red-stained sections, thionin-stained sections were not used in the production of Figs. 81–122. In addition, Sections 58–80 of the first edition have been redrawn from the original luxol fast blue–neutral red sections. Sections to be traced were projected at a known magnification with a projecting microscope. In the case of the coronal sections, only unilateral tracings were made. The drawings were made bilateral by using the mirror image of the unilateral tracing to ensure symmetry. Completed drawings

were then reduced photographically to match the photomicrographs.

Cortical areas are indicated in the schematic drawings with heavy solid lines; myelinated fibers are outlined with thinner solid lines; approximate outlines of cell groups, nuclei, and subcortical areas are indicated by dashed lines; and the cerebellar cortex is outlined with dotted lines.

SYSTEMS OF COORDINATES

It has been our experience that two types of stereotaxic coordinate systems are commonly used with the rat. In the coronal drawings of this atlas we have included both systems for the investigator's convenience.

In both systems the rat's head must be in a fixed position such that the interaural line (an imaginary line that passes through the centers of the ear plugs when the rat is placed in the stereotaxic instrument) is exactly five (5) mm below the level of the upper incisor bar (See Fig. A).

System A. This is the coordinate system of de Groot (1959). In this system the horizontal zero plane is tangent to the upper incisor bar and 5.0 mm above the interaural line (Fig. A). It passes through the anterior and posterior commissures. The dorsal–ventral coordinates (in millimeters) are indicated by the scale to the right of each schematic drawing. The horizontal zero plane of de Groot is indicated by ''0''. The rostral–caudal coordinate (in millimeters) is in the upper right corner of the coronal drawings (Figs. 1–99) and along the

4

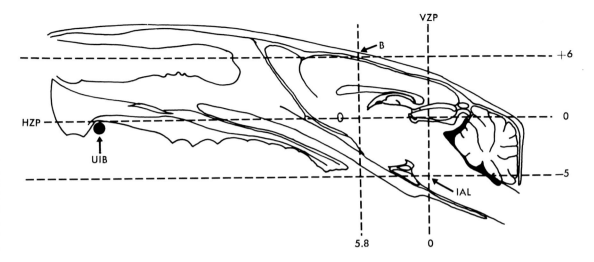

Figure A. Midsagittal section through the head of a rat reconstructed with permission from the atlas of König and Klippel (1963). Broken lines represent planes of the de Groot (1959) coordinate system. B, bregma; HZP, horizontal zero plane; IAL, interaural line; UIB, upper incisor bar; VZP, vertical zero plane.

bottom scale on the sagittal drawings (Figs. 100–122). This coordinate is preceded by a minus (−) sign when it signifies a plane posterior to the vertical zero plane. The medial–lateral coordinate (in millimeters) is obtained from the bottom scale of the coronal drawings and is in the lower right corner on the sagittal drawings.

System B. This system is a simple modification of the de Groot system that is used in conjunction with the coronal drawings only. The rostral–caudal zero reference point is the skull landmark bregma (the point at which the coronal suture crosses the sagittal suture). The rostral–caudal coordinate (in millimeters) for this system is in the upper left corner of each coronal drawing. Sections posterior to bregma are numbered with a minus (−) sign. For those investigators who prefer to

measure the depth of their placements from the dura rather than from stereotaxic zero, the horizontal zero on the left side of each coronal drawing has been placed at the highest point of the cortex or cerebellum.

Determination of Coordinates. If, for example, one wanted to determine the stereotaxic coordinates for the placement of an electrode in the anterior part of the medial geniculate nucleus (GM), the System A coordinates derived from coronal drawing #47 would be as follows: rostral–caudal, 2.8 [that is, 2.8 mm anterior to the vertical zero plane of de Groot (found in the upper right corner of the drawing)]; medial–lateral, 3.0–4.0 mm lateral to the midline (from the scale along the bottom of the drawing); and dorsal–ventral, −0.5–1.5 mm ventral to the horizontal zero plane (from the scale on the right side of the drawing).

5

Using System B the coordinates would be as follows: rostral–caudal, −3.0 [that is, 3.0 mm posterior to bregma (from the upper left corner of the drawing)]; medial–lateral, 3.0–4.0 mm from the midline (from the bottom scale); and dorsal–ventral, 5.75–6.75 mm ventral from the surface of the dura (from the scale on the left side of the drawing).

Stereotaxic coordinates may be obtained in a similar manner from the sagittal sections. In this case, the rostral–caudal coordinate is read from the bottom scale; the medial–lateral coordinate is given in the lower right corner of the drawing; and the dorsal–ventral coordinate, relative to the horizontal zero plane, is obtained from the scale on the right side of the drawing. It is probable that coordinates obtained from the sagittal sections will differ slightly from those obtained from coronal sections. These discrepancies are largely the result of individual differences in brains and are well within the margin of error of stereotaxic techniques.

We have deliberately omitted instructions for the use of stereotaxic instruments, since we feel that adequate coverage of this topic already exists in readily available sources (see, for example, Pellegrino and Cushman, 1971; Sidowski, 1966; Singh and Avery, 1975; Skinner, 1971).

ACKNOWLEDGMENTS

Without the encouragement and support of Dr. Robert B. Malmo, *A Stereotaxic Atlas of the Rat Brain* would never have been produced. And without the equally valuable support of Dr. Joseph Altman, this expanded second edition would have remained only a vain hope. We thank them both for making this work possible.

In addition, we were assisted on the first edition by H. Buchtel, K. Carlson, C. Hodge, R. Musty, A. Pellegrino, and R. S. Snider. Contributions to the second edition were made by S. Bayer, S. Evander, M. Ladd, R. Struble, and C. Walters.

This project was supported in part by grants to R. B. Malmo from the National and Medical Research Councils of Canada and from the National Institutes of Health (U.S. Public Health Service) and the National Science Foundation, and by grants to J. Altman from the National Institutes of Health (U.S. Public Health Service) and the National Science Foundation.

6

REFERENCES

Altman, J. 1969. Autoradiographic and histological studies of postnatal neurogenesis. III. Dating the time of production and onset of differentiation of cerebellar microneurons in rats. *J. Comp. Neurol.* **136:**269–294.

de Groot, J. 1959. The rat forebrain in stereotaxic coordinates. *Verh. K. Ned. Akad. Wet., B. Natuurkd.* **2:**1–40.

Klüver, H., and Barrera, E. 1953. A method for the combined staining of cells and fibers in the nervous system. *J. Neuropathol. Exp. Neurol.* **12:**400–403.

König, J. F. R., and Klippel, R. A. 1963. *The Rat Brain: A Stereotaxic Atlas of the Forebrain and Lower Parts of the Brain Stem.* Williams and Wilkins, Baltimore.

Korneliussen, H. K. 1968. On the morphology and subdivision of the cerebellar nuclei of the rat. *J. Hirnforsch.* **10:**109–122.

Krettek, J. E., and Price, J. L. 1978. Amygdaloid projections to subcortical structures within the basal forebrain and brainstem in the rat and cat. *J. Comp. Neurol.* **178:** 225–253.

Larsell, O. 1937. The cerebellum. *Arch. Neurol. Psychiatry* **38**:580–607.

Lockard, I. 1977. *Desk Reference For Neuroanatomy: A Guide to Essential Terms.* Springer-Verlag, New York.

Pellegrino, L. J., and Cushman, A. J. 1971. Use of the stereotaxic technique. In: *Methods in Psychobiology,* Vol. I (R. D. Myers, ed.). Academic Press, New York.

Sidman, R. L., Angevine, J. B., Jr., and Pierce, E. T. 1971. *Atlas of the Mouse Brain and Spinal Cord.* Harvard University Press, Cambridge.

Sidowski, J. B. 1966. *Experimental Methods and Instrumentation in Psychology.* McGraw-Hill, New York.

Singh, D., and Avery, D. D. 1975. *Physiological Techniques in Behavioral Research.* Brooks Cole, Monterey.

Skinner, J. E. 1971. *Neuroscience: A Laboratory Manual.* Saunders, Philadelphia.

Slotnick, B. M., and Leonard, C. M. 1975. *A Stereotaxic Atlas of the Albino Mouse Forebrain.* Publication No. (ADM) 75-100. U.S. Department of Health, Education and Welfare, Rockville, Md.

Whishaw, I. Q., Cioe, J. D., Previsich, N., and Kolb, B. 1977. The variability of the interaural line vs. the stability of bregma in rat stereotaxic surgery. *Physiol. Behav.* **19**:719–722.

Wünscher, W., Schober, W., and Werner, J. 1965. *Architectonischer Atlas Vom Hirnstamm Der Ratte.* S. Hirzel, Leipzig.

Zeman, W., and Innes, J. R. M. 1963. *Craigie's Neuroanatomy of the Rat.* Academic Press, New York.

Index of Abbreviations

Abb.	Structure	Fig. no.	Abb.	Structure	Fig. no.
A	Aqueduct of Sylvius; Aqueductus cerebri (Sylvii)	49–69, 100–102	ACO	Cortical amygdaloid nucleus; Nucleus amygdaloideus corticalis	30–44, 114–121
AAA	Anterior amygdaloid area; Area amygdaloidea anterior	20–29, 113–122	AD	Anterodorsal nucleus of the thalamus; Nucleus anterodorsalis thalami	32–36, 105–107
ABL	Basal amygdaloid nucleus, lateral part; Nucleus amygdaloideus basalis, pars lateralis	28–43, 118–122	AHA	Anterior hypothalamic area; Area anterior hypothalami	26–31, 100–105
			AHP	Amygdalo-hippocampal area	117–122
ABM	Basal amygdaloid nucleus, medial part; Nucleus amygdaloideus basalis, pars medialis	33–37, 117–121	AL	Lateral amygdaloid nucleus; Nucleus amygdaloideus lateralis	27–43, 121–122
ACB	Lateral parolfactorial area; Nucleus accumbens septi; Area parolfactoria lateralis	10–25, 103–112	ALV	Alveus	115–122
			AM	Anteromedial nucleus of the thalamus; Nucleus anteromedialis thalami	30–34, 100–104
ACE	Central amygdaloid nucleus; Nucleus amygdaloideus centralis	28–37, 117–120	AMB	Nucleus ambiguus	88–91, 108–109
			AME	Medial amygdaloid nucleus; Nucleus amygdaloideus medialis	30–38, 114–119

9

FA	Amygdaloid fissure; Fissura amygdaloidea	44, 47–48, 50
f.ap.	Ansoparamedian fissure; Fissura ansoparamediana	93–96
FC	Cuneate fasciculus; Fasciculus cuneatus	86–99, 105–111
FCS	Corticospinal fibers; Fibrae corticospinalis	97–99, 100–101
FD	Dentate gyrus; Fascia dentata	36–60, 101–121
FG	Fasciculus gracilis	94–99, 100–106
FH	Hippocampal fissure; Fissura hippocampi	43–59, 105–116
FI	Fimbria of the hippocampus; Fimbria hippocampi	33–46, 107–122
FL	Fornix longus (of Forel)	28–30
fl.	Flocculus	70–80, 118–121
FLD	Dorsal fasciculus of Schütz; Dorsal longitudinal bundle; Fasciculus longitudinalis dorsalis (Schütz)	41–57
FLM	Medial longitudinal fasciculus; Fasciculus longitudinalis medialis	51–99, 100–107
FP	Pontine fibers; Fibrae pontis	58–65, 100–112
f. pfl.	Parafloccular fissure; Fissura parafloccularis	75–83
f. pl.	Posterolateral fissure; Fissura posterolateralis	88–92, 100–106
f. ppd.	Prepyramidal fissure; Fissura prepyramidalis	86–97
f. pr.	Fissura prima	72–84, 100–113
f. p. s.	Posterior superior fissure; Fissura posterioris superioris	75–92
FR	Rhinal fissure; Fissura rhinalis	1–59, 108–111
FS	Nucleus fimbrialis septi	103
f. sec.	Fissura secunda	91–99, 100–107
FTN	Trigeminal nerve fibers; Fibrae nervi trigemini	65–70, 111–112
FUD	Dorsal funiculus; Funiculus dorsalis	101–104
FX	Fornix	26–42, 100–106
GCC	Genu of the corpus callosum; Genu corporis callosi	101–110
GE	Nucleus gelatinosus thalami	35–39, 101–103
GLD	Lateral geniculate body, dorsal part; Corpus geniculatum laterale, pars dorsalis	43–50, 115–119
GLV	Lateral geniculate body, ventral part; Corpus geniculatum laterale, pars ventralis	41–48, 116–119
GM	Medial geniculate body; Corpus geniculatum mediale	47–57, 115–118

13

MCL	Mitral cell layer of the olfactory bulb; Lamina cellularum mitralium bulbi ofactorii	103–112	NCD	Dorsal cochlear nucleus; Nucleus cochlearis dorsalis	77–83, 111–118
MD	Dorsomedial nucleus of the thalamus; Nucleus mediodorsalis thalami	33–42, 101–107	NCL	Nucleus centralis lateralis thalami	109–112
MFB	Median forebrain bundle; Fasciculus medialis telencephali	10–44, 105–117	NCP	Bed nucleus of the posterior commissure; Nucleus proprius commissurae posterioris	49–57
ML	Lateral mamillary nucleus; Nucleus mamillaris lateralis	37–44, 104–105	NCV	Ventral cochlear nucleus; Nucleus cochlearis ventralis	68–79, 116–119
MM	Medial mamillary nucleus; Nucleus mamillaris medialis	40–42, 100–102	ND	Dentate nucleus; Lateral deep nucleus of the cerebellum; Nucleus dentatus	77–83, 114–118
MP	Posterior mamillary nucleus; Nucleus mamillaris posterior	40–44, 100–103	NF	Fastigial nucleus; Medial deep nucleus of the cerebellum; Nucleus fastigius	79–86, 103–107
MPA	Medial parolfactorial area; Area parolfactoria medialis	7–19, 100–102	NI	Interpositus nucleus; Intermediate deep nucleus of the cerebellum; Nucleus interpositus	78–80, 107, 113–116
MPO	Medial preoptic area; area praeoptica medialis	20–25, 100–101			
MS	Medial septal nucleus; Nucleus medialis septi	18–25, 100–103			
MT	Mamillothalamic tract; Bundle of Vicq d'Azyr; Tractus mamillothalamicus	32–41, 100–104	NIa	Interpositus nucleus, anterior part; Intermediate deep nucleus of the cerebellum, anterior part; Nucleus interpositus, pars anterior	81–83, 108–112
MTT	Mesencephalic tract of the trigeminal (V) nerve; Tractus mesencephalici nervi trigemini	66–68, 106–109			

15

OA	Anterior olfactory nucleus; Nucleus olfactorius anterior	7–11, 104, 106–109, 112–118
OAD	Anterior olfactory nucleus, dorsal part; Nucleus olfactorius anterior, pars dorsalis	1–6, 107–111
OAE	Anterior olfactory nucleus, external part; Nucleus olfactorius anterior, pars externa	1–4, 102–112
OAL	Anterior olfactory nucleus, lateral part; Nucleus olfactorius anterior, pars lateralis	1–6, 107–112
OAM	Anterior olfactory nucleus, medial part; Nucleus olfactorius anterior, pars medialis	1–6, 102–106
OL	Inferior olivary nuclei; Nuclei olivarii inferior	78–92, 100–107
OS	Superior olivary nucleus; Nucleus olivaris superior	66–69
OT	Optic tract; Tractus opticus	28–49, 102–119
P	Pons	51–59, 100–111
PAM	Periamygdaloid cortex	116–122
PAS	Parasubiculum	118–119
PBL	Lateral parabrachial nucleus; Nucleus parabrachialis lateralis	67–68, 109–113

PBM	Medial parabrachial nucleus; Nucleus parabrachialis medialis	65–69, 108–111
PC	Cerebral peduncle; Pedunculus cerebri	36–57, 107–117
PCI	Inferior cerebellar peduncle; Pedunculus cerebellaris inferior	75–91, 112–117
PCM	Middle cerebellar peduncle; Pedunculus cerebellaris medius	61–72, 113–119
PF	Parafascicular nucleus of the thalamus; Nucleus parafascicularis thalami	41–48, 102–107
pfl.	Paraflocculus	73–87, 120–122
PH	Posterior nucleus of the hypothalamus; Nucleus posterior hypothalami	35–40, 101–103
PIR	Piriform cortex; Cortex piriformis	7–44, 114–122
PL	Paralemniscal nucleus; Nucleus paralemniscalis	57–63, 114–115
PM	Mamillary peduncle; Pedunculus mamillaris	41–48, 102–105
PMD	Dorsal premamillary nucleus; Nucleus praemamillaris dorsalis	37–40, 100–102
PMV	Ventral premamillary nucleus; Nucleus praemamillaris ventralis	35–39, 102–106

SC	Suprachiasmatic nucleus; Nucleus suprachiasmaticus	22–26, 100–103
SCC	Splenium of the corpus callosum; Splenium corporis callosi	103–110, 117–122
SEL	Subependymal layer of the olfactory bulb; Lamina subependymalis bulbi olfactorii	108
SGM	Stratum griseum mediale colliculi superioris	100–113
SGP	Stratum griseum profundum colliculi superioris	100–106
SGS	Stratum griseum superficiale colliculi superioris	100–112
s. i. c.	Intercrural sulcus; Sulcus intercruralis	85–93, 115–116
SM	Stria medullaris thalami	26–46, 100–111
SN	Substantia nigra	43–56, 105–115
SO	Supraoptic nucleus of the hypothalamus; Nucleus supraopticus hypothalami	23–29, 107–111
SOC	Superior olivary complex	105–113
SOL	Solitary nucleus; Nucleus solitarius	80–99, 100–110
SOP	Stratum opticum colliculi superioris	100–113
ST	Stria terminalis; Taenia semicircularis	26–42, 109–120
STM	Stria medullaris of the fourth ventricle; Stria medullares ventriculi quarti	80–82, 113–114
SUM	Supramamillary nucleus; Nucleus supramamillaris	41–44, 100–104
TCL	Lateral corticospinal tract; Tractus corticospinalis lateralis	97–98
TCS	Corticospinal tract; Tractus corticospinalis	56–99, 100–107
TIC	Olfacto-cortical tract; Tractus olfacto-corticalis	103–104
TL	Lateral tegmental nucleus; Nucleus lateralis tegmenti	51–56
TOC	Olivo-cerebellar tract; Tractus olivocerebellaris	85–93, 100–102
TOI	Intermediate olfactory tract; Tractus olfactorius intermedius	1–6, 104–112
TOL	Lateral olfactory tract; Tractus olfactorius lateralis	1–25, 107–120
TOM	Medial olfactory tract; Tractus olfactorius medialis	1–6

TP	Tuberculopiriform tract; Tractus tuberculopiriformis	10–19
TPO	Tegmental nucleus of the pons; Nucleus tegmenti pontis	58–64, 100–105
TRS	Rubrospinal tract; Tractus rubrospinalis	65–88, 107–114
TS	Nucleus triangularis septi	26–30, 102–103
TSC	Spinocerebellar tract; Tractus spinocerebellaris	67–69, 79–99, 108–117
TSL	Solitary tract; Tractus solitarius	80–98, 103–113
TST	Spinal tract of the trigeminal (V) nerve; Tractus spinalis nervi trigemini	72–99, 111–117
TSTH	Spinothalamic tract; Tractus spinothalamicus	77–99, 106–108
TT	Mamillotegmental tract; Tractus mamillotegmentalis	39–49
TTS	Tectospinal tract; Tractus tectospinalis	58–67, 100–103
TUO	Olfactory tubercle; Tuberculum olfactorium	7–21, 102–115
V	Ventricle; Ventriculus	8–57, 70–80, 100–122

VA	Ventral nucleus of the thalamus, anterior part; Nucleus ventralis thalami, pars anterior	32–36, 105–115
VB	Ventral nucleus of the thalamus, basal part; Nucleus ventralis thalami, pars basalis	104–118
VD	Ventral nucleus of the thalamus, dorsal part; Nucleus ventralis thalami, pars dorsalis	37–45
VE	Ventral nucleus of the thalamus; Nucleus ventralis thalami	33–49
VM	Ventral nucleus of the thalamus, medial part; Nucleus ventralis thalami, pars medialis	35–39, 104–106
VMH	Ventromedial nucleus of the hypothalamus; Nucleus ventromedialis hypothalami	28–36, 100–104
VO	Olfactory ventricle; Ventriculus olfactorius	1–7, 107
VTN	Ventral tegmental nucleus of Tsai; Nucleus ventralis tegmenti (Tsai)	45–50, 102–111
ZI	Zona incerta	30–46, 103–116
ZT	Transitional zone of the amygdala; Zona transitionalis	36–40

Index of Structures

Structure	Abb.	Fig. no.
Acoustic nerve	VIII	70–71, 112–118
Alveus	ALV	115–122
Ammon's horn	HPC	7–58, 100–122
Amygdala		
Basal nucleus, lateral part	ABL	28–43, 118–122
Basal nucleus, medial part	ABM	33–37, 117–121
Central nucleus	ACE	28–37, 117–120
Cortical nucleus	ACO	30–44, 114–121
Intercalated nucleus	ICL	25–34
Lateral nucleus	AL	27–43, 121–122
Medial nucleus	AME	30–38, 114–119
Transitional zone	ZT	36–40
Amygdaloid area, anterior	AAA	20–29, 113–122
Amygdaloid fissure	FA	44, 47–48, 50
Ansoparamedian fissure	f. ap.	93–96
Anterior commissure	CA	7–25
Anterior part	CAa	100–112
Bed nucleus	BCA	22–25, 101–102
Posterior part	CAp	100–118

Structure	Abb.	Fig. no.
Aquaeductus cerebri (Sylvii)	A	49–69, 100–102
Arcuate nucleus of the medulla	ARC	100
Area		
Amygdalo-hippocampal	AHP	117–122
Anterior amygdaloid	AAA	20–29, 113–122
Anterior hypothalamic	AHA	26–31, 100–105
Lateral hypothalamic	LHA	26–40, 105–110
Lateral parolfactorial	ACB	10–25, 103–112
Lateral preoptic	POA	16–25, 102–115
Medial parolfactorial	MPA	7–19, 100–102
Medial preoptic	MPO	20–25, 100–101
Postrema	AP	94–96
Pretectal	PRT	45–52, 100–111
Brachium conjunctivum	BC	58–73, 102–112
Decussation	DBC	55–57, 61–62, 100–101

Nucleus (*cont.*)

Endopiriformis	ENP	117–122
Entopeduncularis	EP	32–37, 113–117
Fastigius	NF	79–86, 103–107
Fimbrialis septi	FS	103
Gelatinosus thalami	GE	35–39, 101–103
Gracilis	GR	94–99, 100–105
Habenularis lateralis	HL	38–46, 102–104
Habenularis medialis	HM	38–47, 100–102
Hypothalami		
Arcuatus	ARH	26–40, 100–103
Dorsomedialis	DMH	32–36, 100–105
Paraventricularis	PVH	26–31, 100–102
Posterior	PH	35–40, 101–103
Supraopticus	SO	23–29, 107–111
Ventromedialis	VMH	28–36, 100–104
Intermediate (cerebellum)	NI	78–80, 107, 113–116
Anterior part	NIa	81–83, 108–112
Posterior part	NIp	81–86, 108–112
Interpeduncularis	IP	47–58, 100–102
Interpositus	NI	78–80, 107, 113–116
Pars anterior	NIa	81–83, 108–112
Pars posterior	NIp	81–86, 108–112
Lateral (cerebellum)	ND	77–83, 114–118
Lateralis septi	LS	16–29, 101–106
Lateralis tegmenti	TL	51–56
Lateralis thalami	LT	35–45, 105–117
Pars posterior	LTP	46–50, 107–115
Lemnisci lateralis, dorsalis	LLD	59–66, 111–115

Nucleus (*cont.*)

Lemnisci lateralis, ventralis	LLV	59–65, 109–112
of Luys	CL	38–42, 109–113
Mamillaris		
Lateralis	ML	37–44, 104–105
Medialis	MM	40–42, 100–102
Posterior	MP	40–44, 100–103
Medial (cerebellum)	NF	79–86, 103–107
Medialis septi	MS	18–25, 100–103
Mediodorsalis thalami	MD	33–42, 101–107
Nervi		
Facialis	VII	71–78, 107–113
Hypoglossi	XII	89–97, 100–102
Oculomotorii	III	49–60, 62–64, 100–103
Edinger-Westphal	E-W	100–101
Trigemini	V	58–60, 104
Motorius	NTM	67–70, 107–111
Sensorius	NST	65–73, 113–114
Tractus mesencephalici	NMT	67–72, 105–108
Tractus spinalis	NTST	71–99, 109–116
Pars caudalis	NTSTc	113–114
Trochlearis	IV	65–66, 103–104
Vagi	X	92–97, 101–103
Olfactorius anterior	OA	7–11, 104, 106–109, 112–118
Pars dorsalis	OAD	1–6, 107–111
Pars externa	OAE	1–4, 102–112
Pars lateralis	OAL	1–6, 107–112
Pars medialis	OAM	1–6, 102–106

29

Nucleus (*cont.*)

Olivaris inferior	OL	78–92, 100–107
Olivaris superior	OS	66–69
Accessorius	AOS	65–69
Parabrachialis lateralis	PBL	67–68, 109–113
Parabrachialis medialis	PBM	65–69, 108–111
Parafascicularis thalami	PF	41–48, 102–107
Paralemniscalis	PL	57–63, 114–115
Paraolivaris superior	POS	67–69
Parataenialis thalami	PT	28–36, 100–102
Paraventricularis hypothalami	PVH	26–31, 100–102
Paraventricularis thalami	PV	28–46, 100–103
Posterior hypothalami	PH	35–40, 101–103
Posterior thalami	NPT	43–50, 105–113
Premamillaris dorsalis	PMD	37–40, 100–102
Premamillaris ventralis	PMV	35–39, 102–106
Prepositus hypoglossi	NPH	74–89, 100–102
Proprius commissurae anterioris	BCA	22–25, 101–102
Proprius commissurae posterioris	NCP	49–57
Proprius striae terminalis	BST	24–27, 105–108
Raphe	RA	71–76, 101–104
Raphe dorsalis	DR	60–65, 100–102
Raphe pontis	RAP	100–102
Reticularis lateralis magnocellularis	RLM	88–95, 105–111
Reticularis lateralis parvocellularis	RLP	89–90, 112–113

Nucleus (*cont.*)

Reticularis thalami	RT	30–40, 104–119
Retrofacialis	NRF	78–83, 109
Reuniens thalami	RE	28–42, 100–104
Rhomboideus thalami	RH	33–40, 100–103
Ruber	NR	50–55, 101–106
Septi		
Accumbens	ACB	20–25, 103–112
Fimbrialis	FS	103
Lateralis	LS	16–29, 101–106
Medialis	MS	18–25, 100–103
Solitarius	SOL	80–99, 100–110
Striae terminalis, proprius	BST	24–27, 105–108
Subthalamicus (Luys)	CL	38–42, 109–113
Suprachiasmaticus	SC	22–26, 100–103
Supramamillaris	SUM	41–44, 100–104
Supraopticus hypothalami	SO	23–29, 107–111
Tegmenti		
Centralis	CT	58–60
Dorsalis	NTD	67–68, 102
Lateralis	TL	51–56
Pontis	TPO	58–64, 100–105
Ventralis (Tsai)	VTN	45–50, 102–111
Thalami		
Anterodorsalis	AD	32–36, 105–107
Anteromedialis	AM	30–34, 100–104
Anteroventralis	AV	28–36, 106–111
Centralis lateralis	NCL	109–112
Centromedianus	CMT	100–103
Gelatinosus	GE	35–39, 101–103

31

33

35

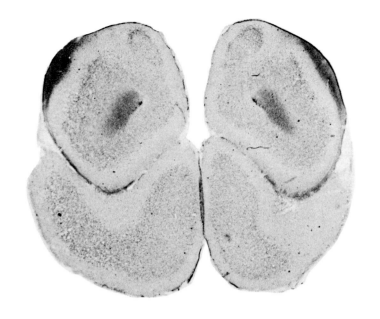

L

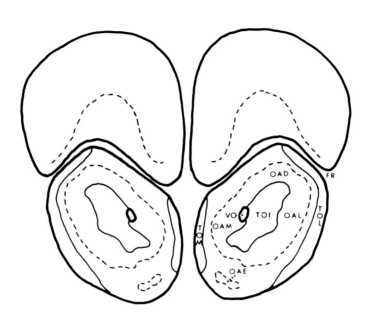

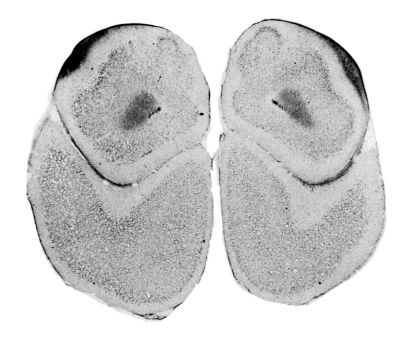

2

6.0

11.8

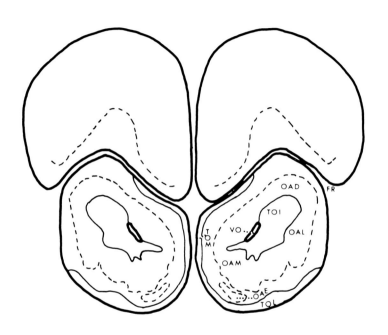

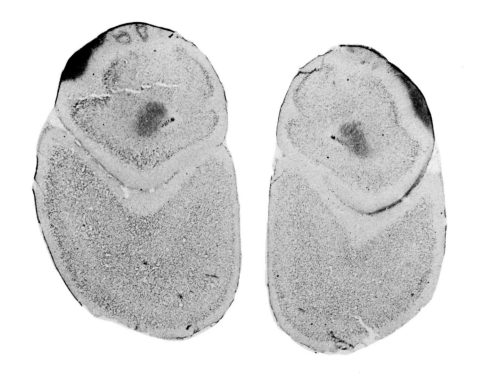

3

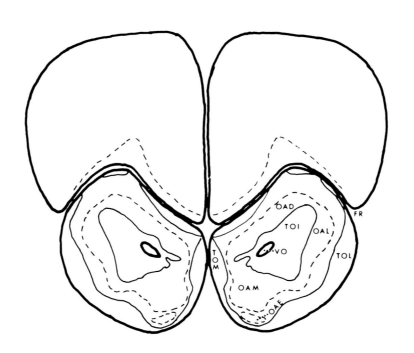

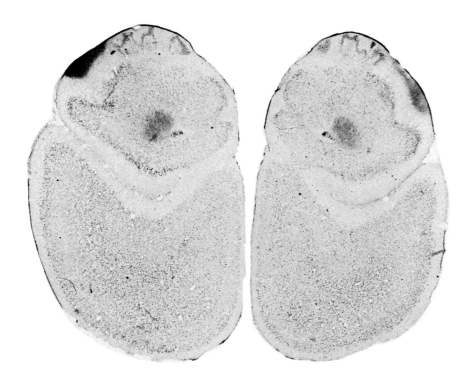

4

5.6

11.4

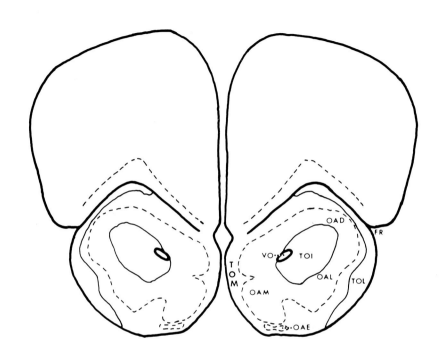

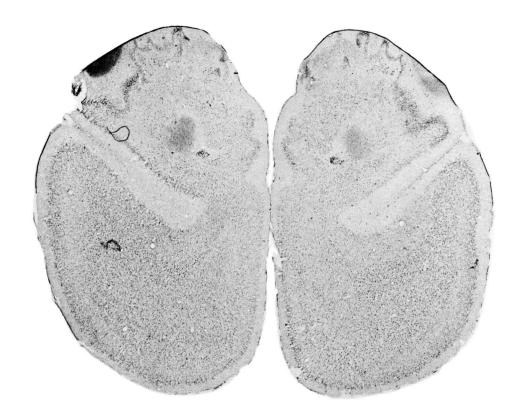

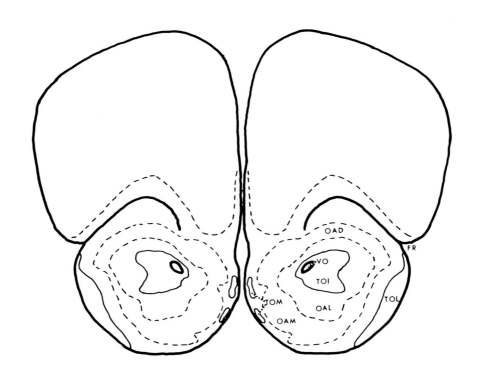

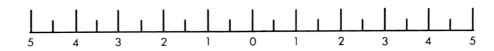

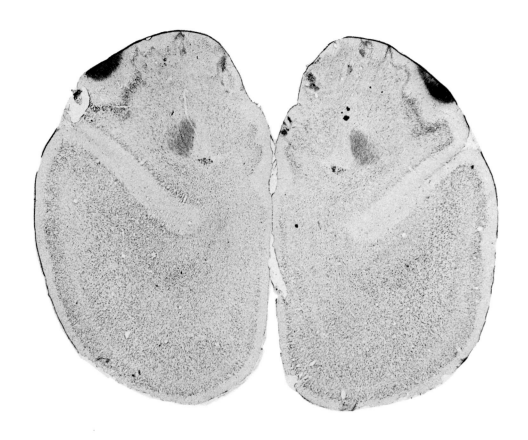

9

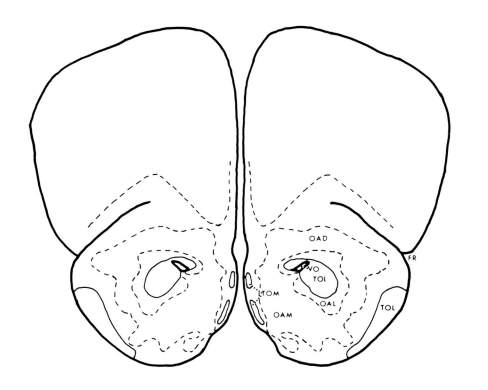

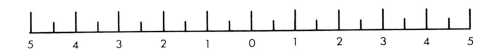

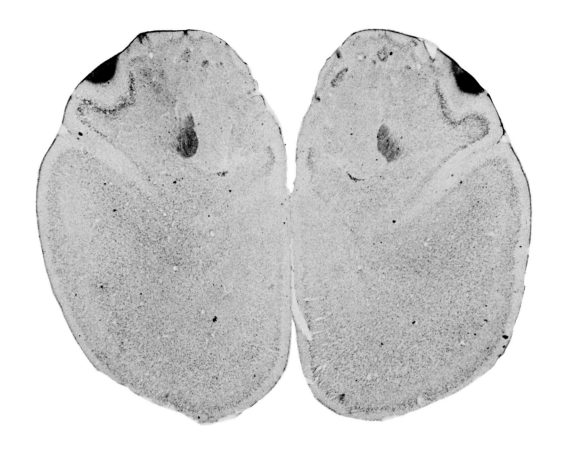

7

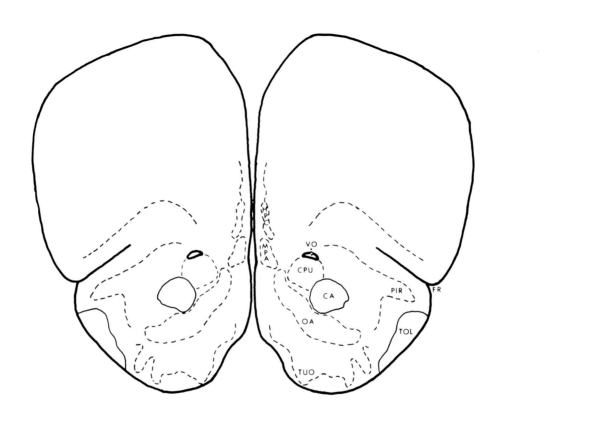

5.0
10.8

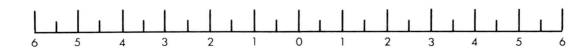

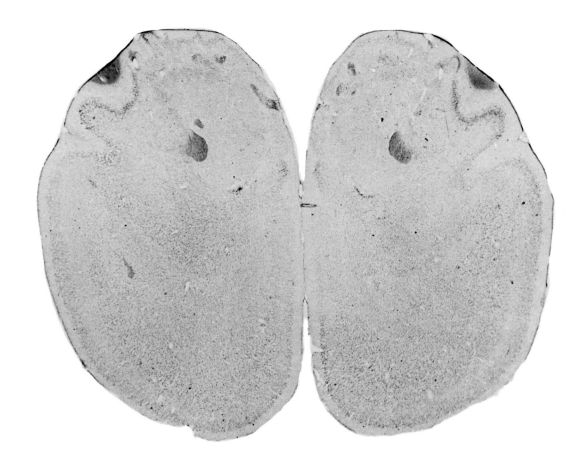

8

4.8　　　　　　　　　　　　　　　　　10.6

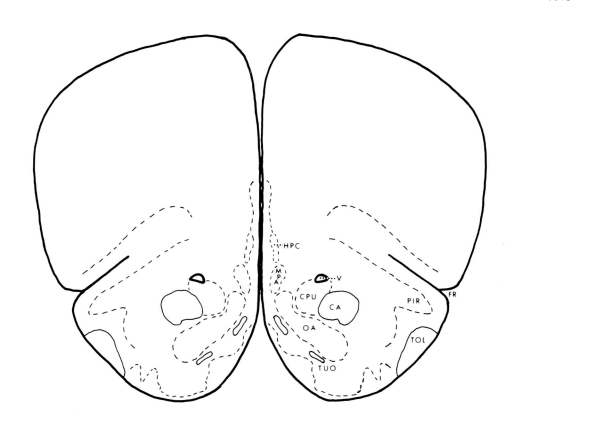

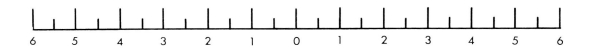

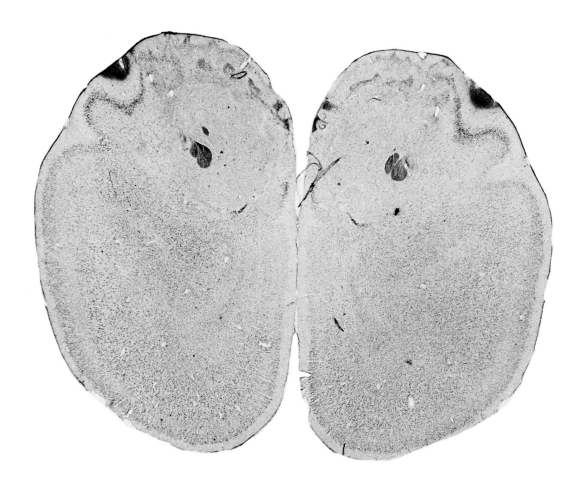

6

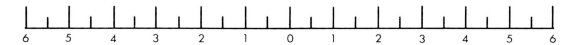

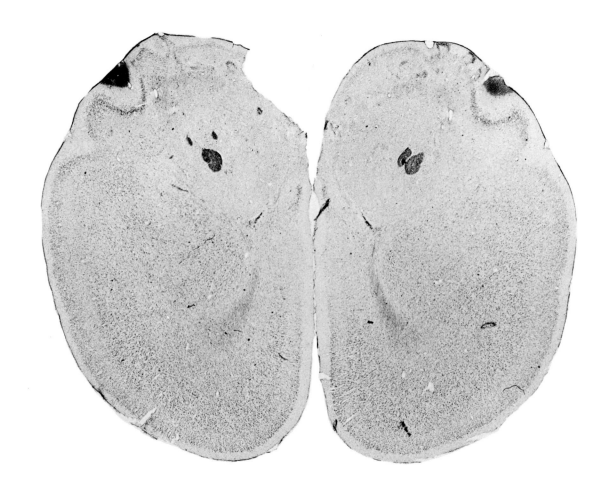

10

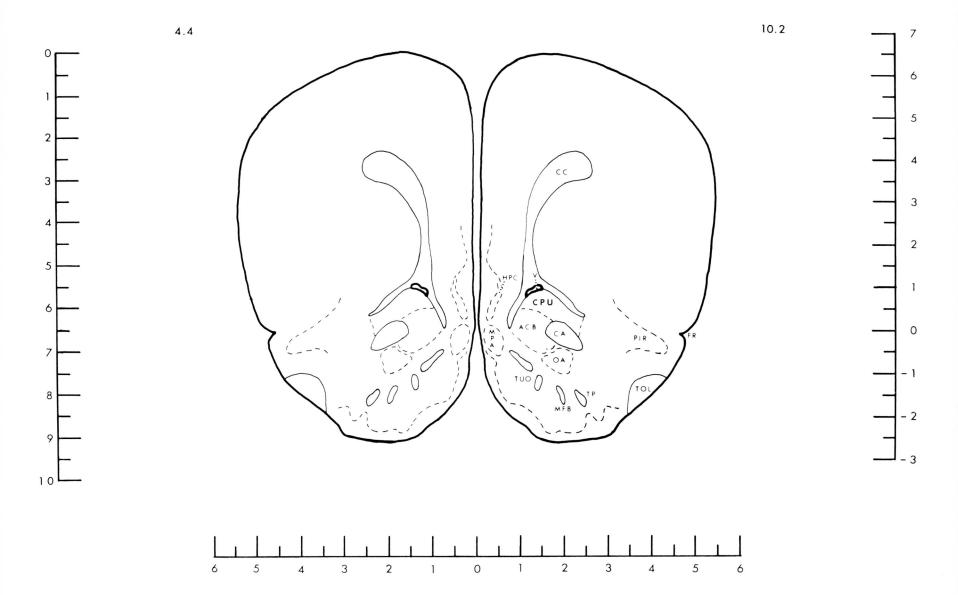

4.4

10.2

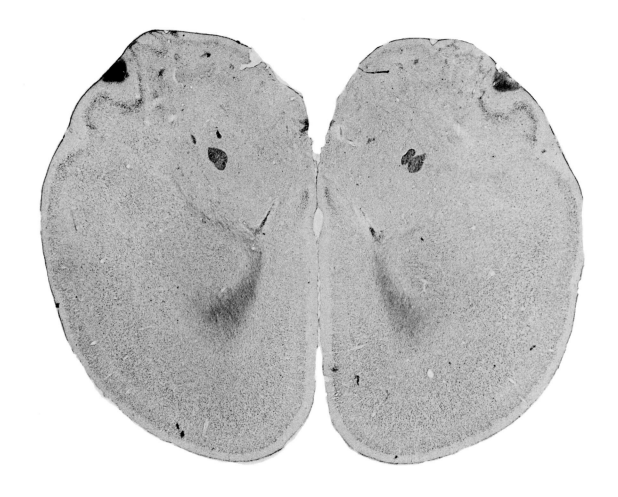

77

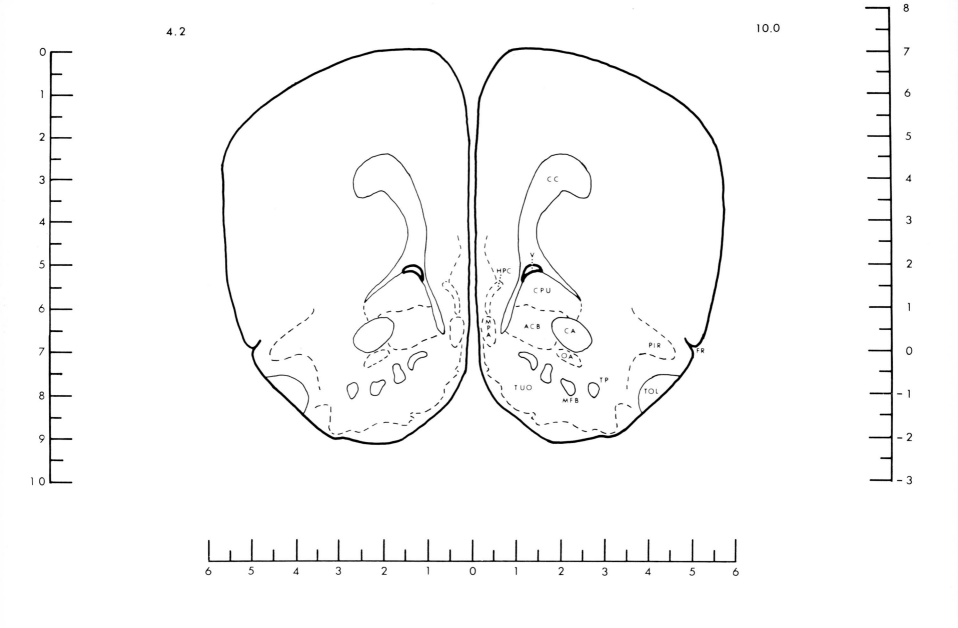

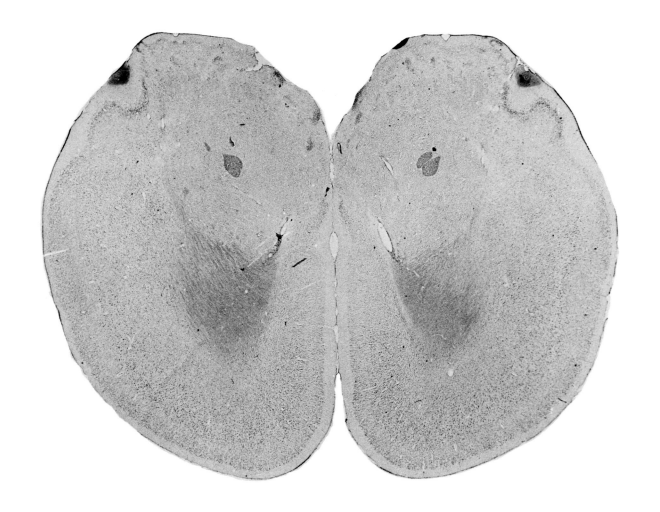

12

4.0   9.8

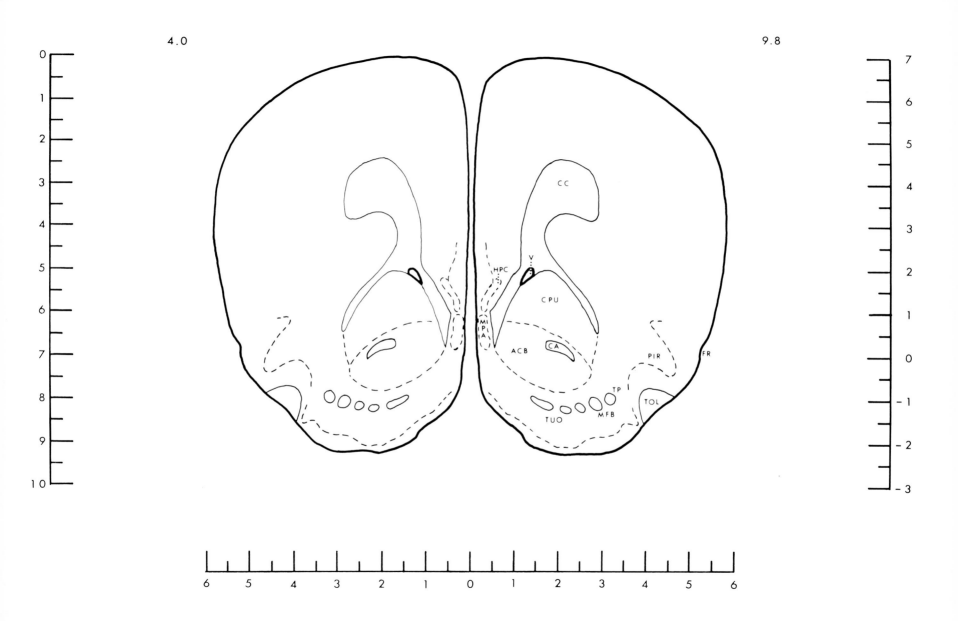

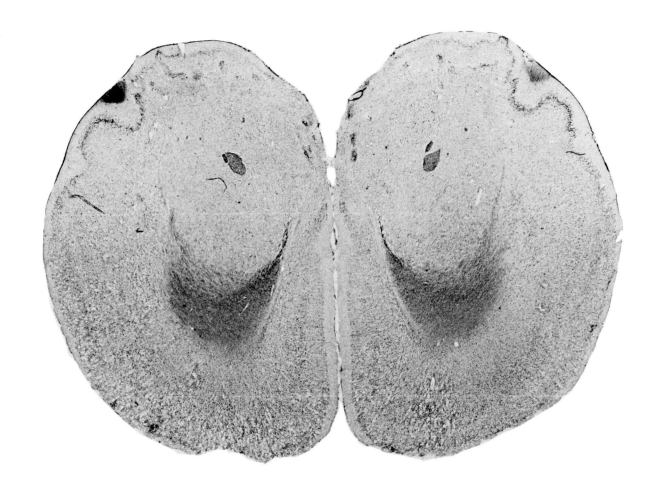

13

3.8   9.6

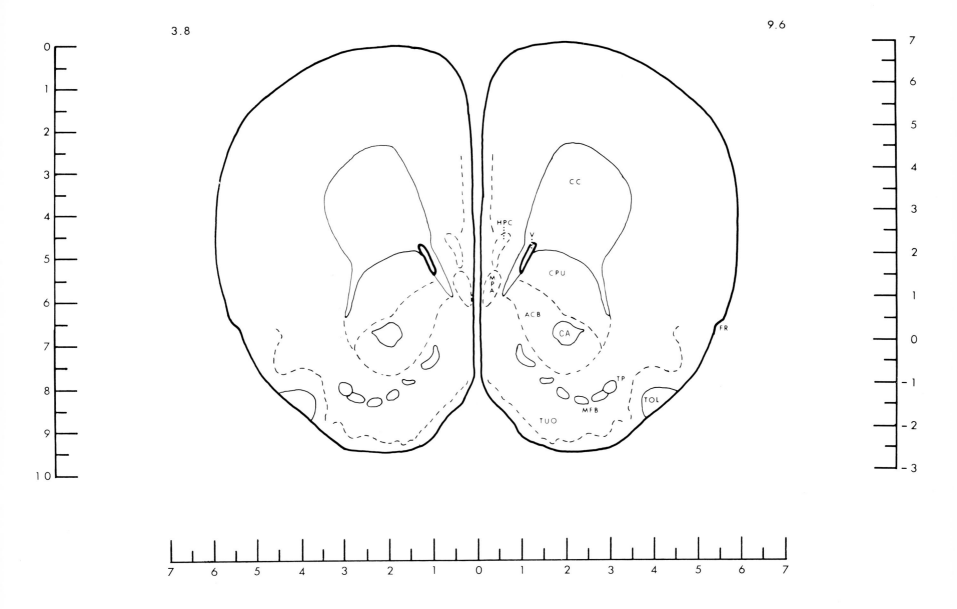

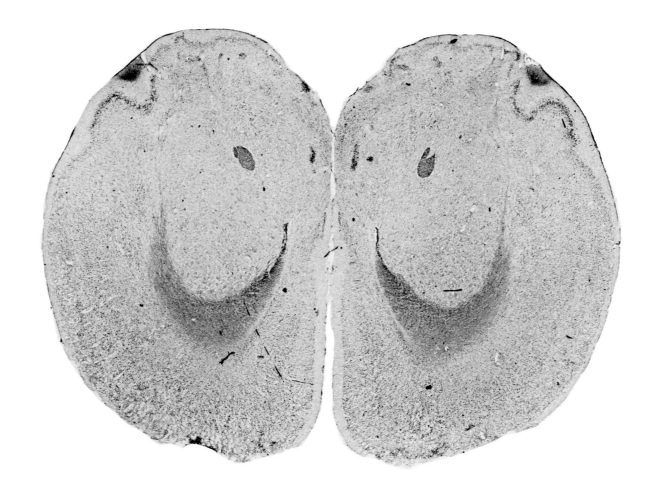

14

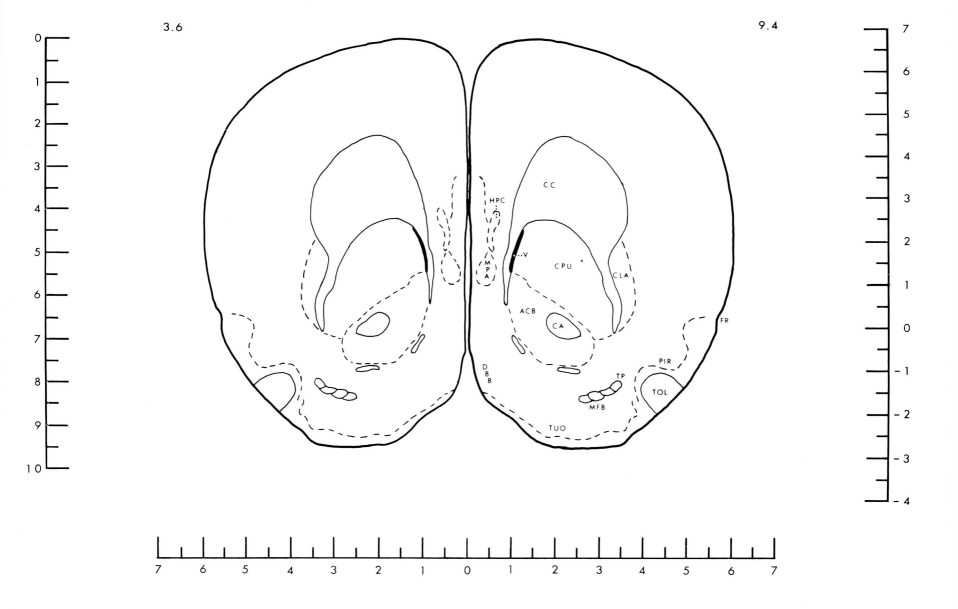

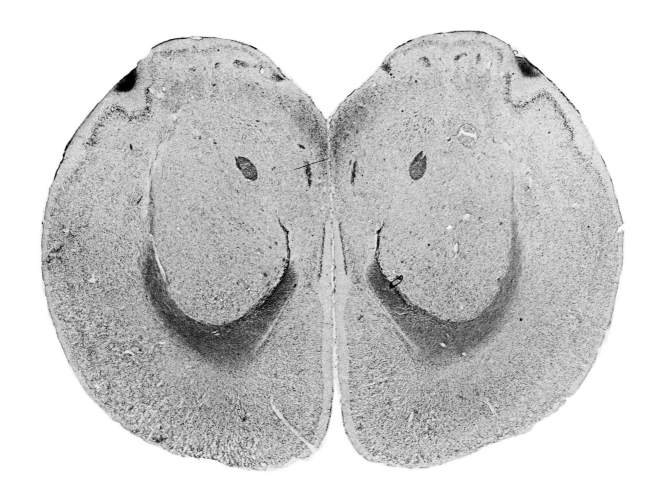

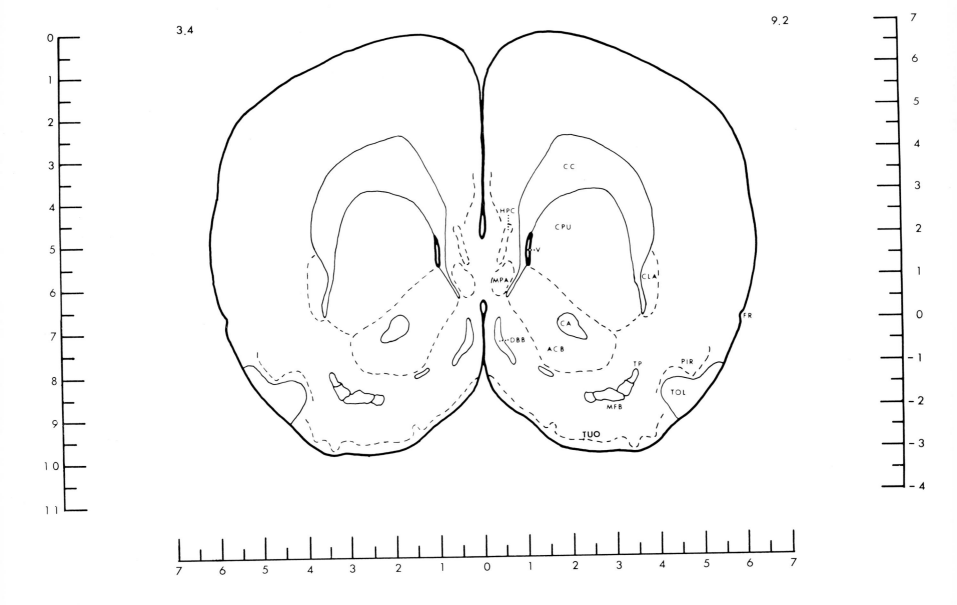

3.4

9.2

CC

HPC

CPU

I.V

MPA

CLA

CA

DBB

ACB

TP

PIR

FR

MFB

TOL

TUO

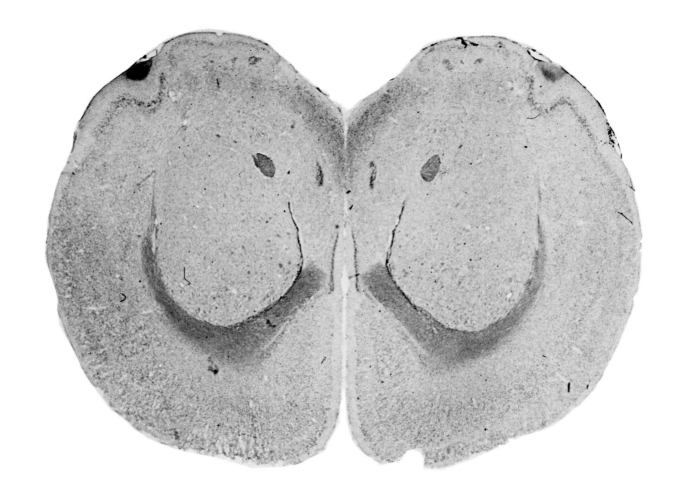

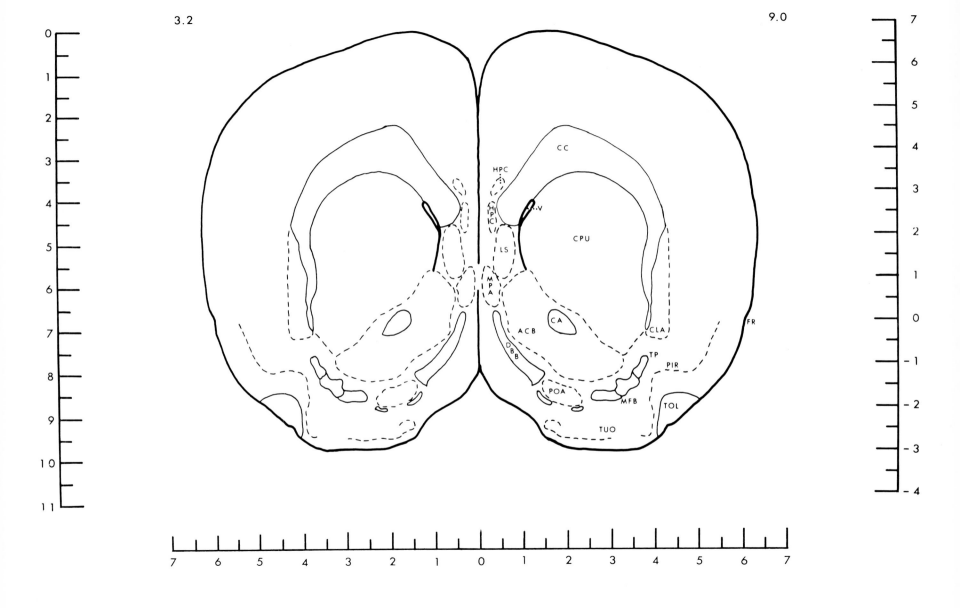

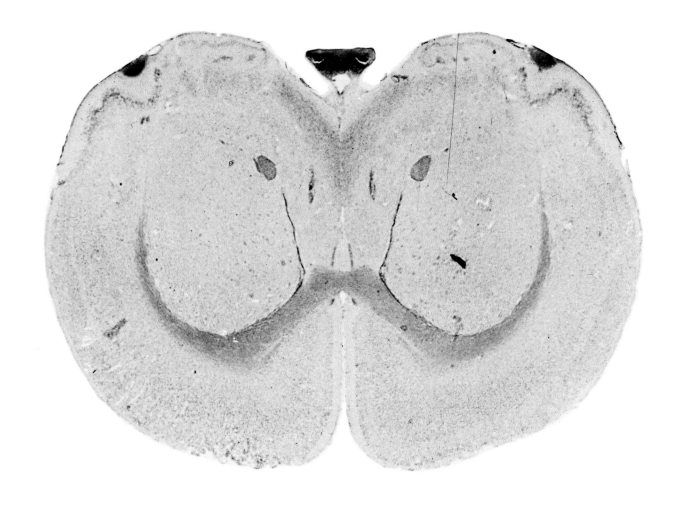

17

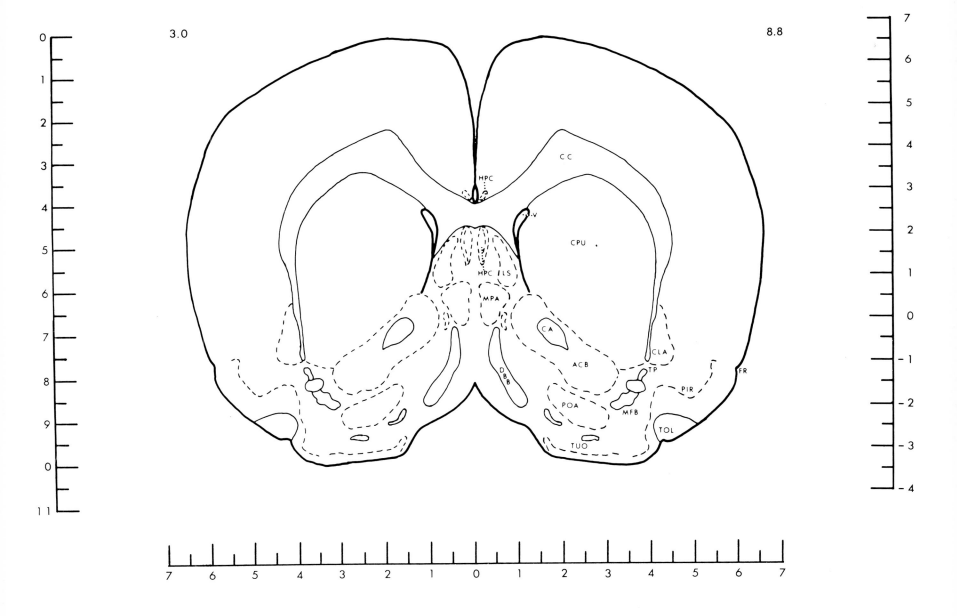

3.0 8.8

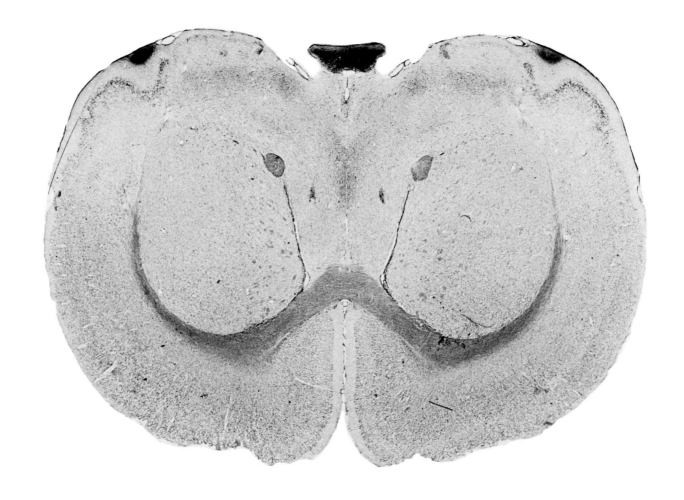

18

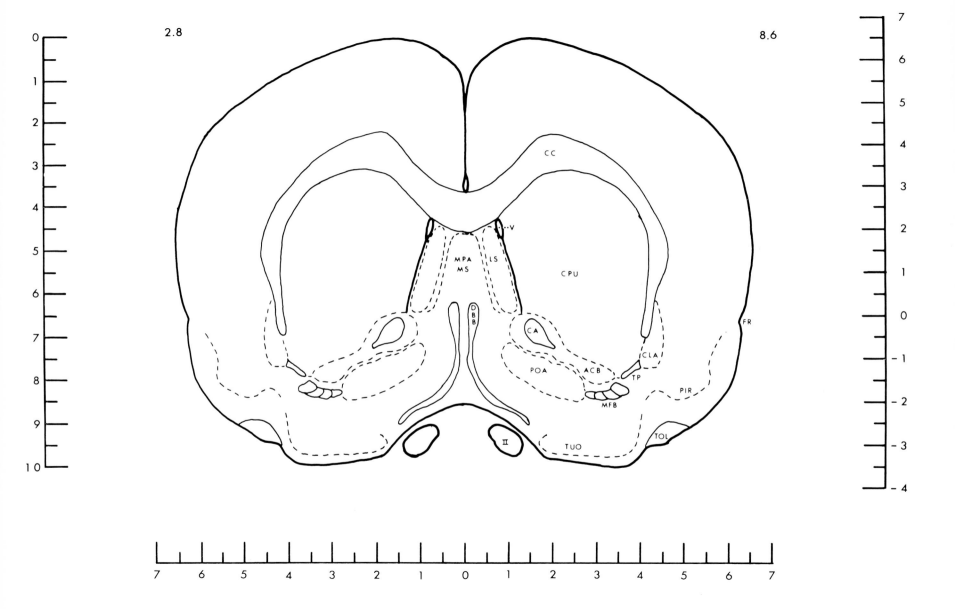

2.8　　　　　　　　　　　　　　　　　　　　　8.6

CC

V

MPA
MS

LS

CPU

D
B
B

CA

FR

CLA

POA

ACB

TP

PIR

MFB

II

TUO

TOL

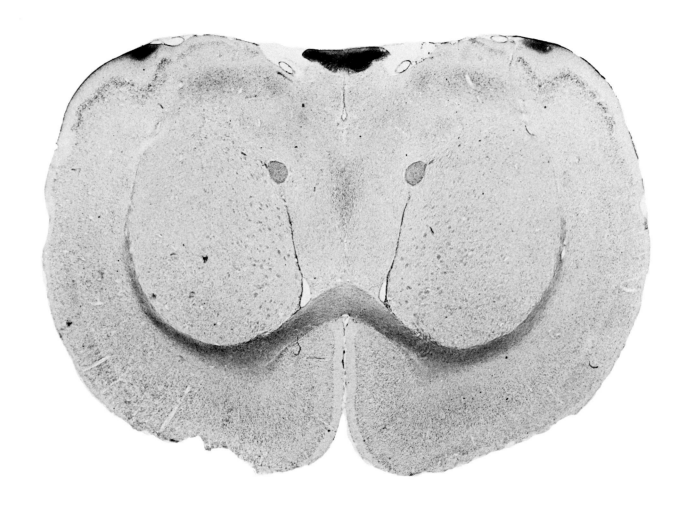

61

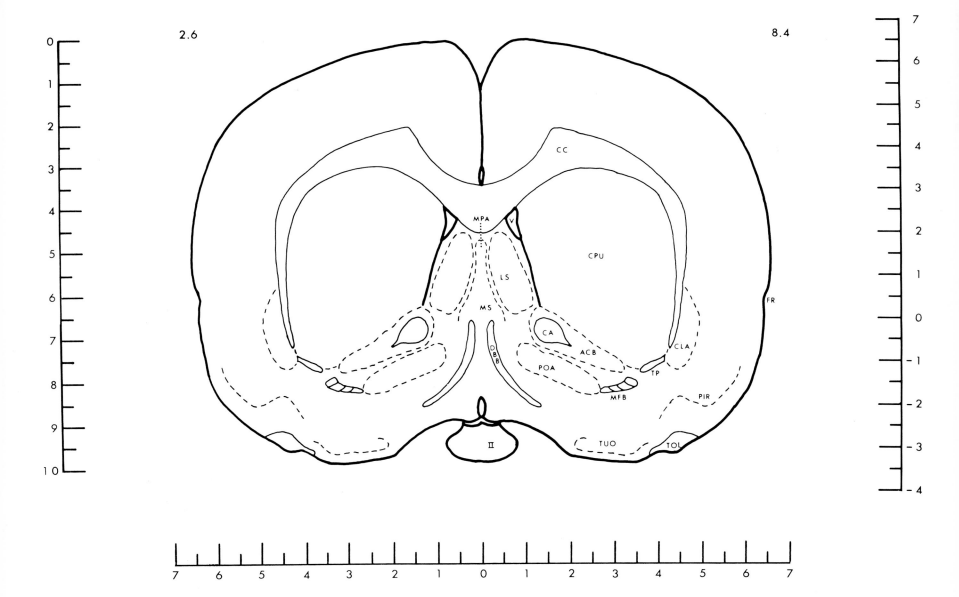

2.6   8.4

CC

MPA

V

CPU

LS

MS

CA

DBB

ACB

CLA

POA

TP

MFB

PIR

FR

II

TUO

TOL

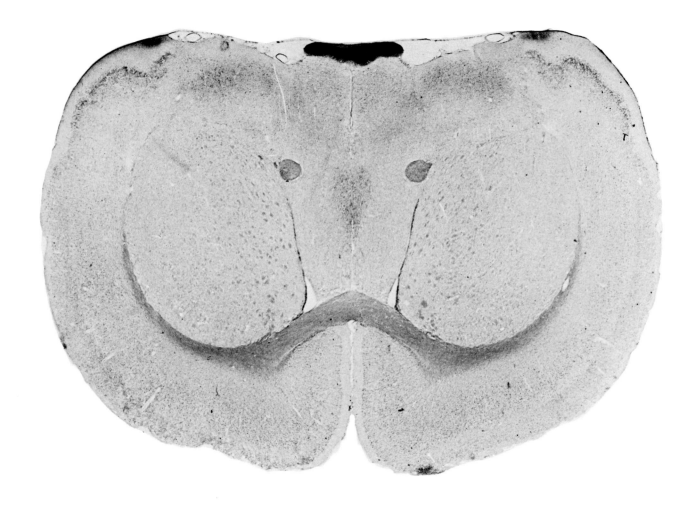

20

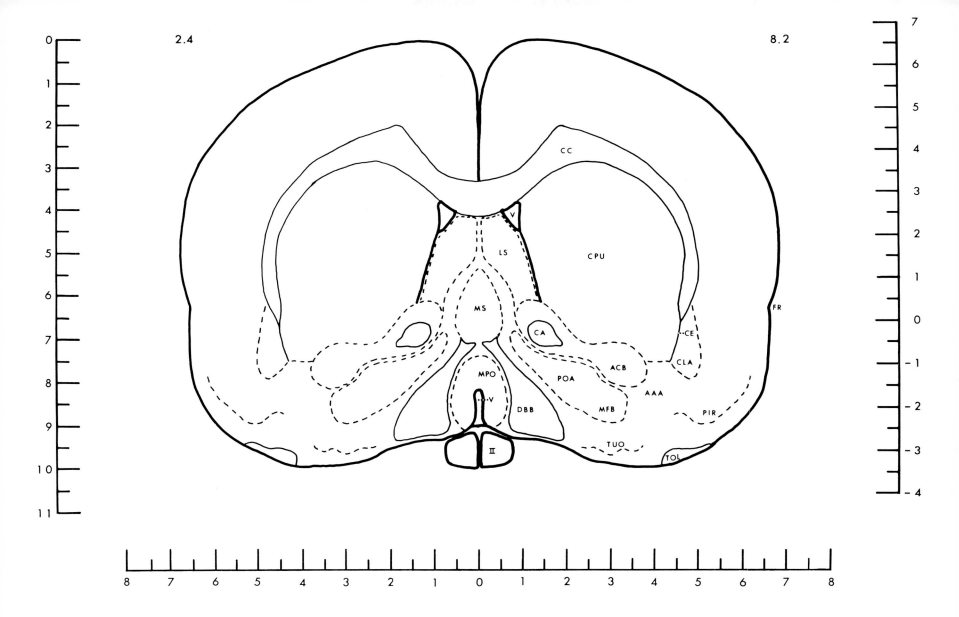

2.4

8.2

CC

V

LS

CPU

MS

CA

CE

ACB

CLA

FR

MPO

POA

AAA

V

DBB

MFB

PIR

II

TUO

TOL

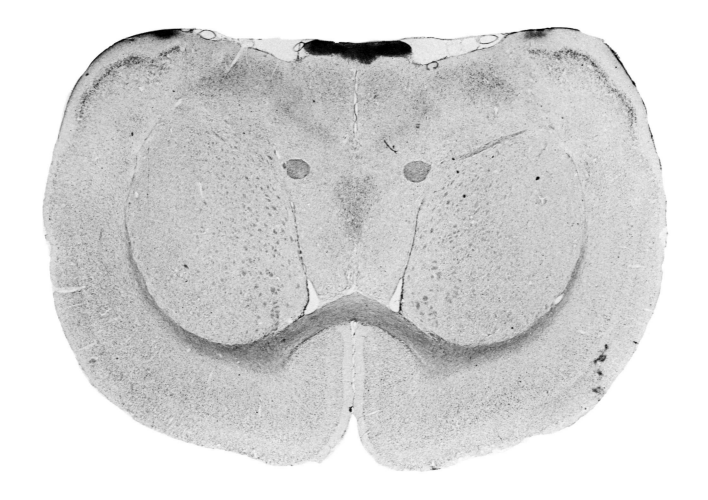

21

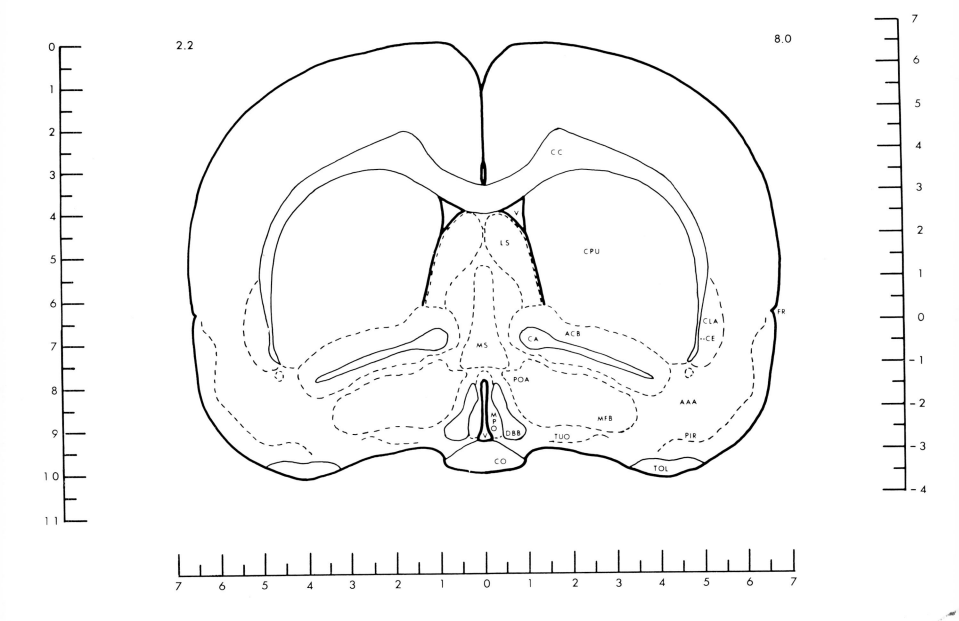

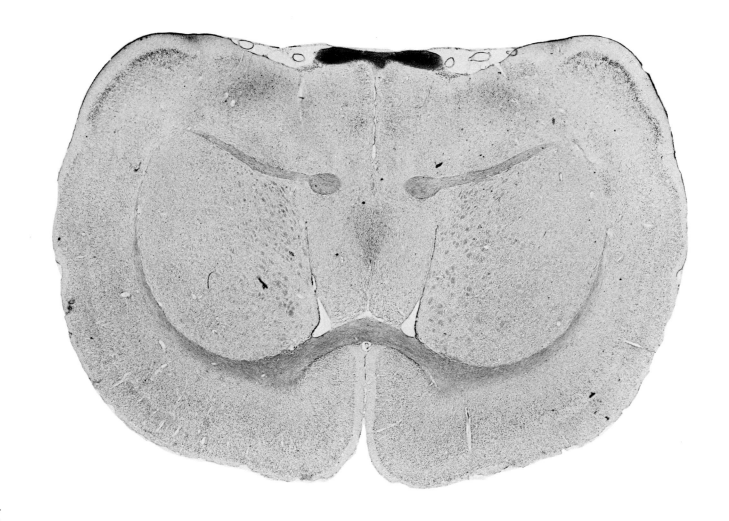

22

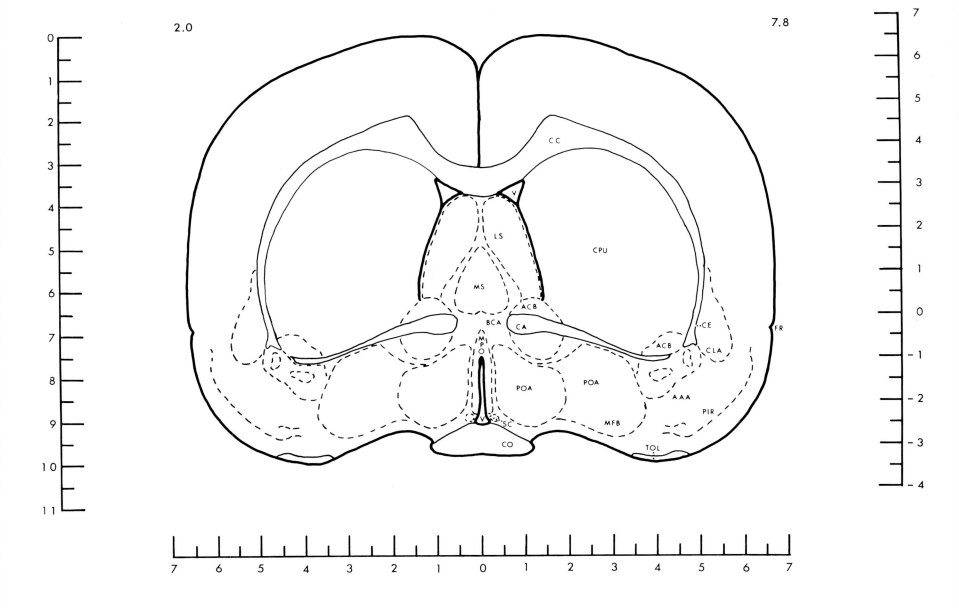

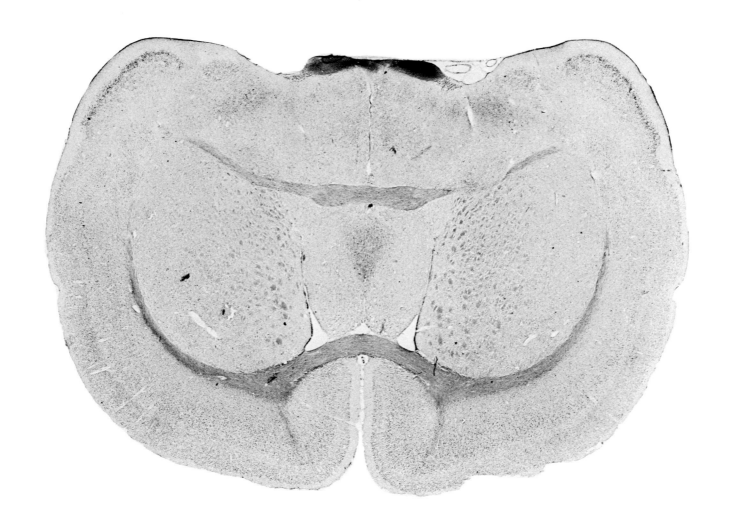

23

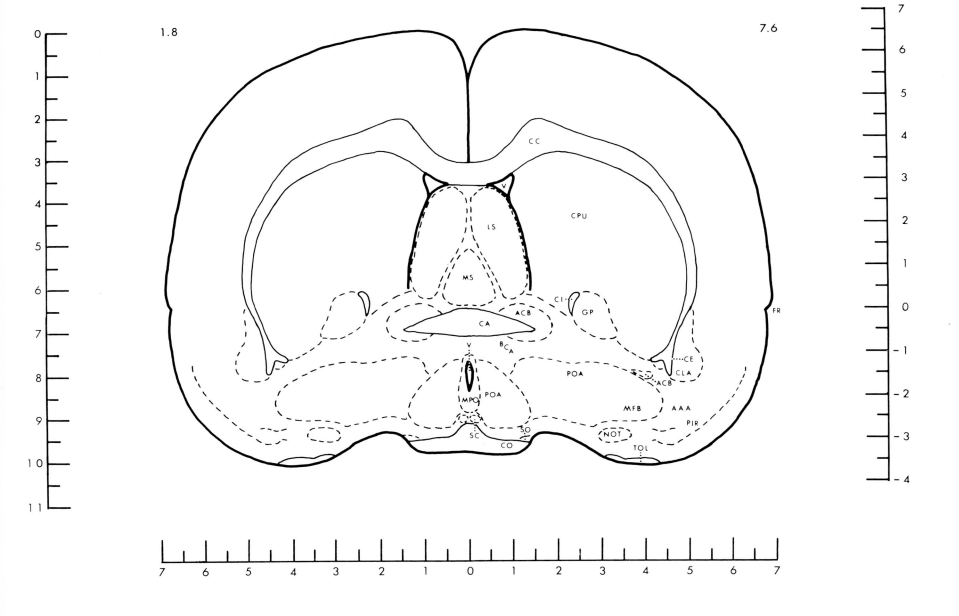

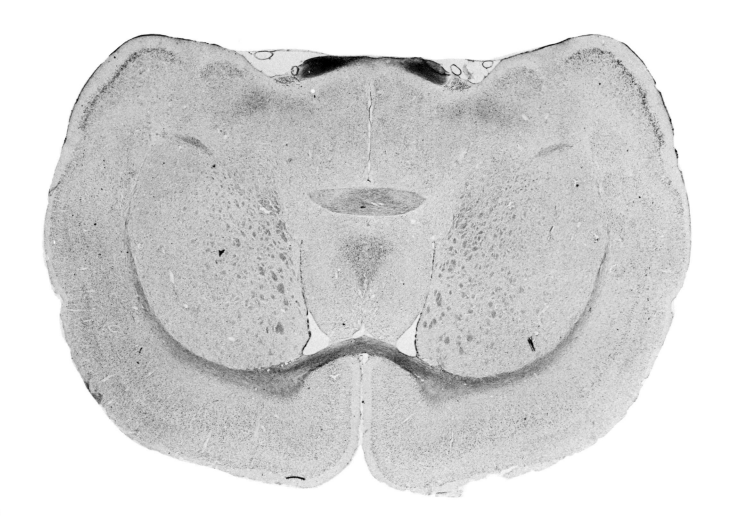

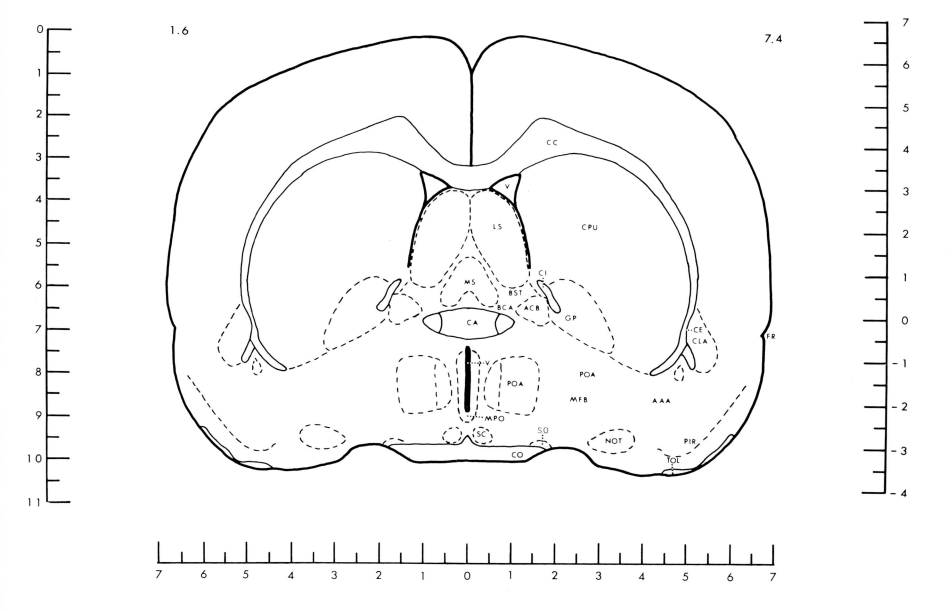

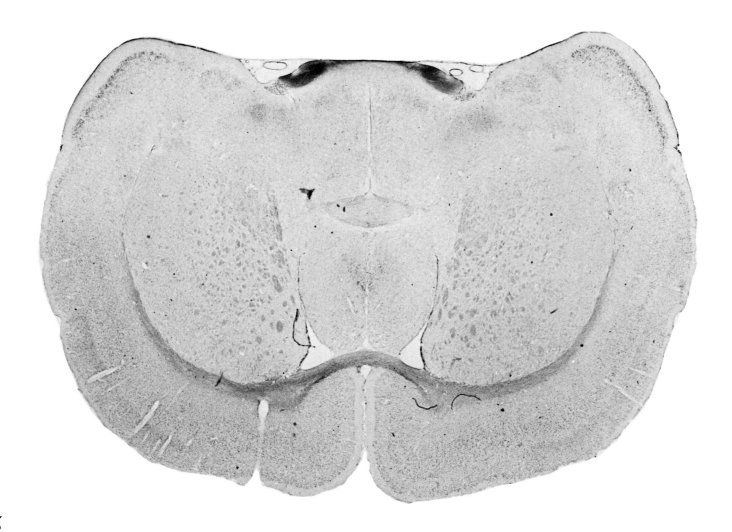

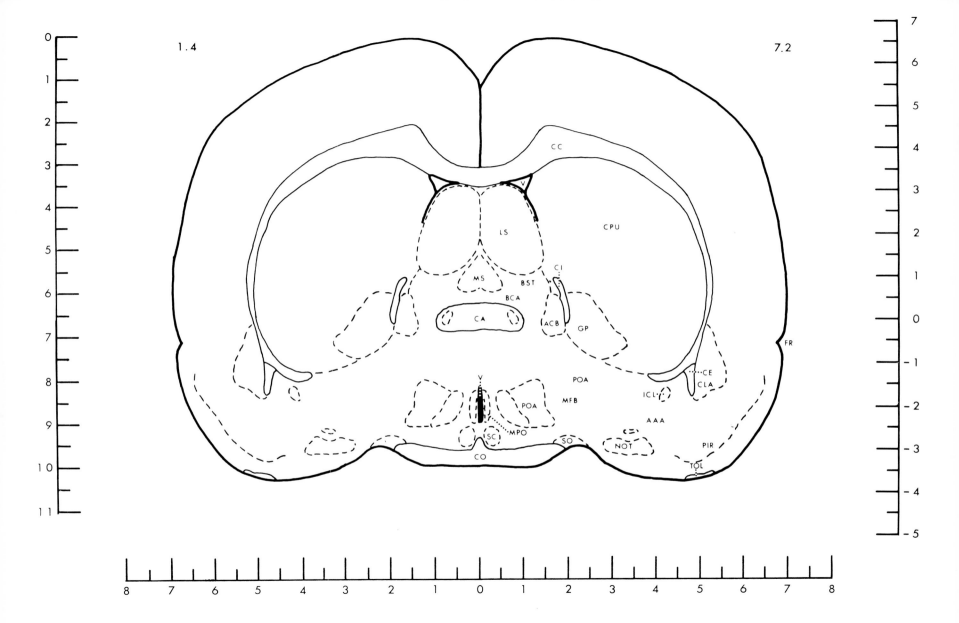

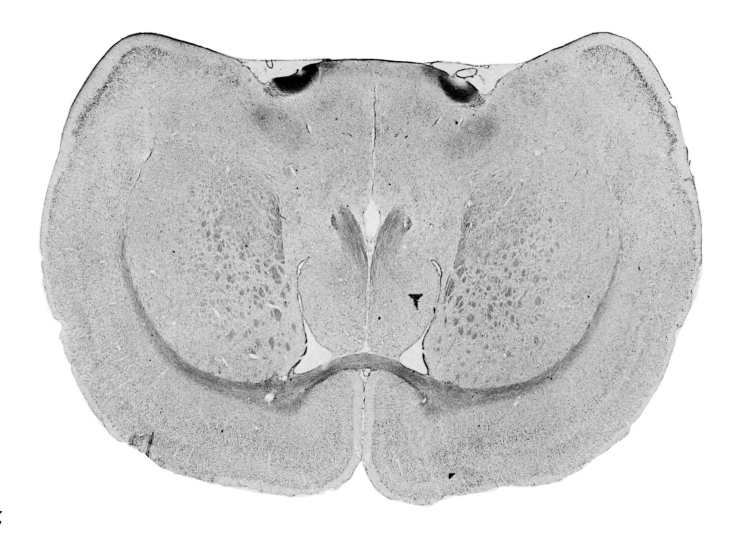

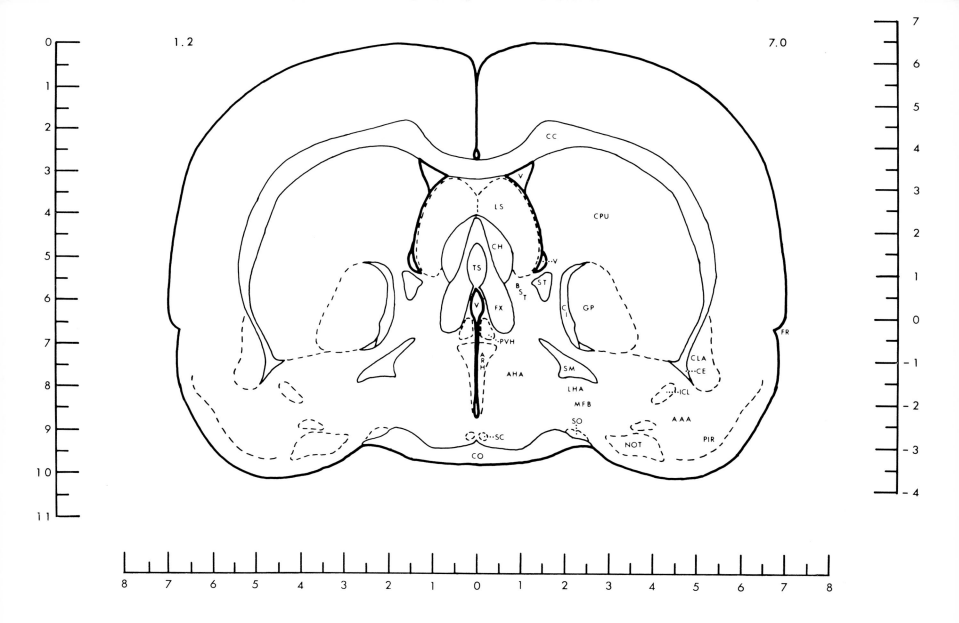

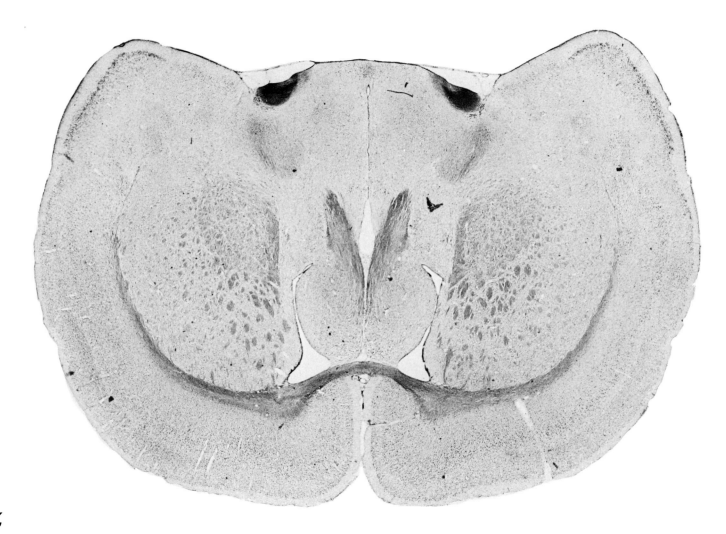

27

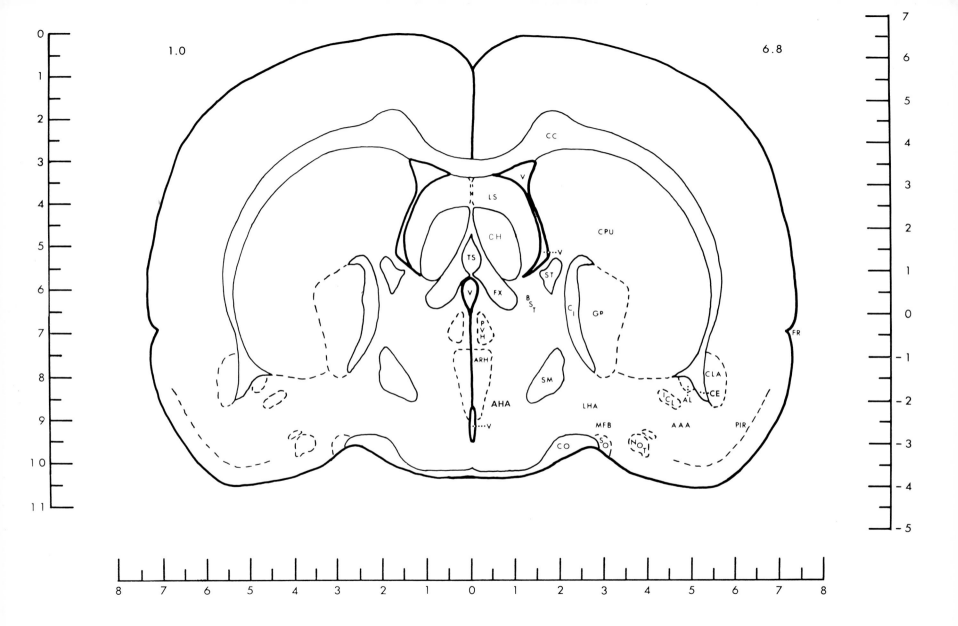

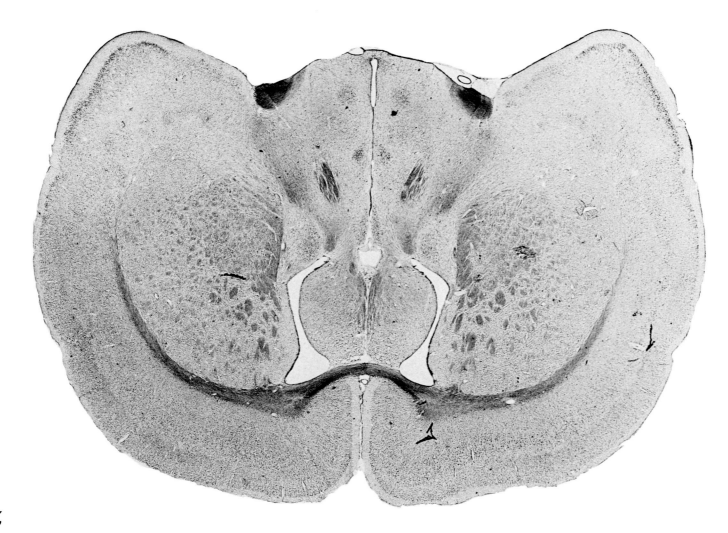

28

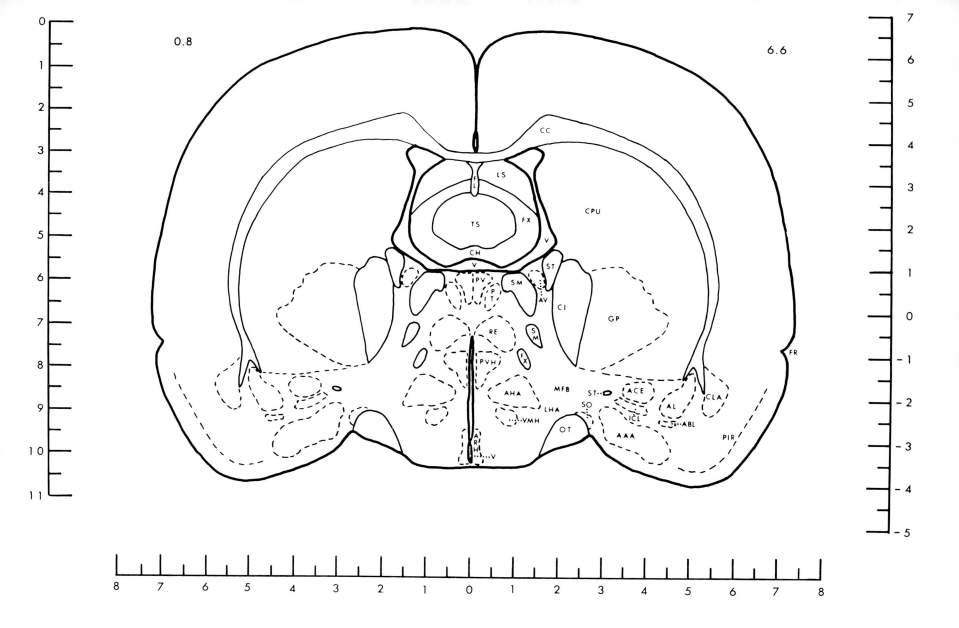

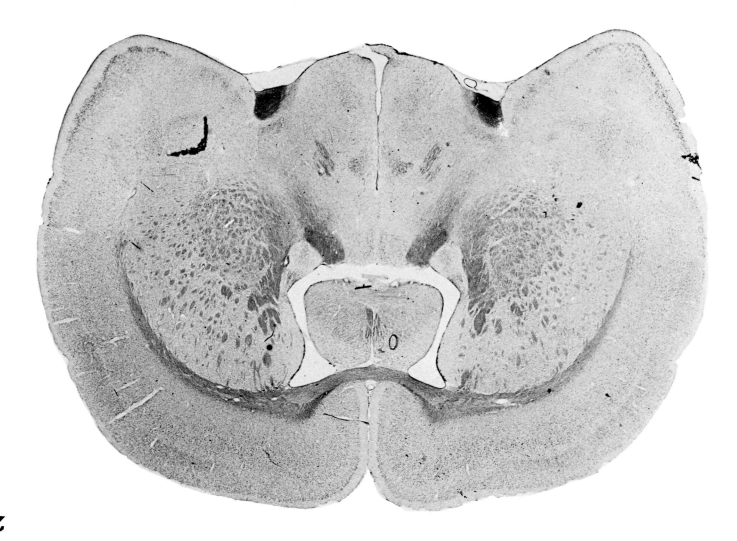

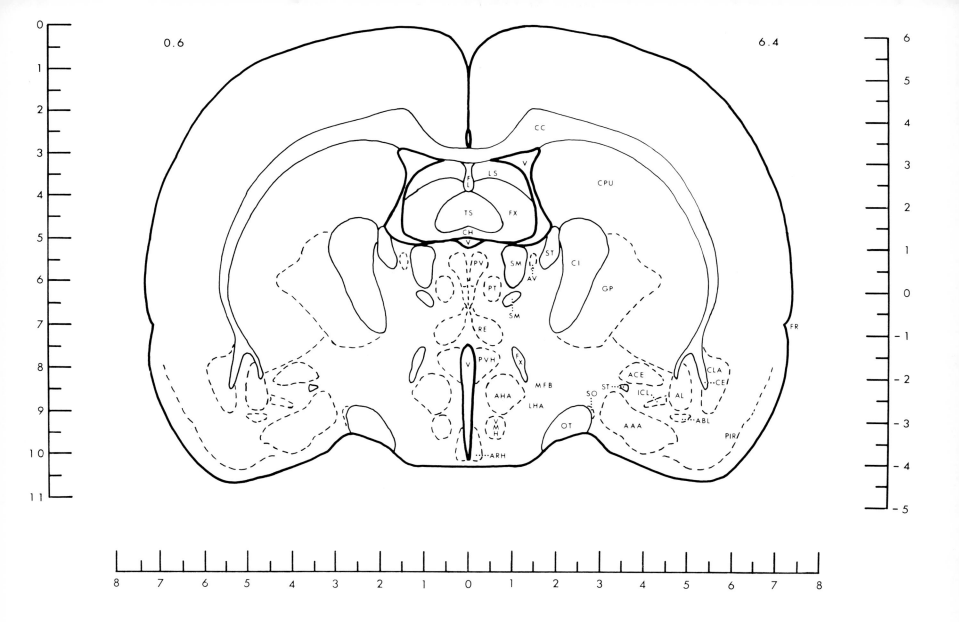

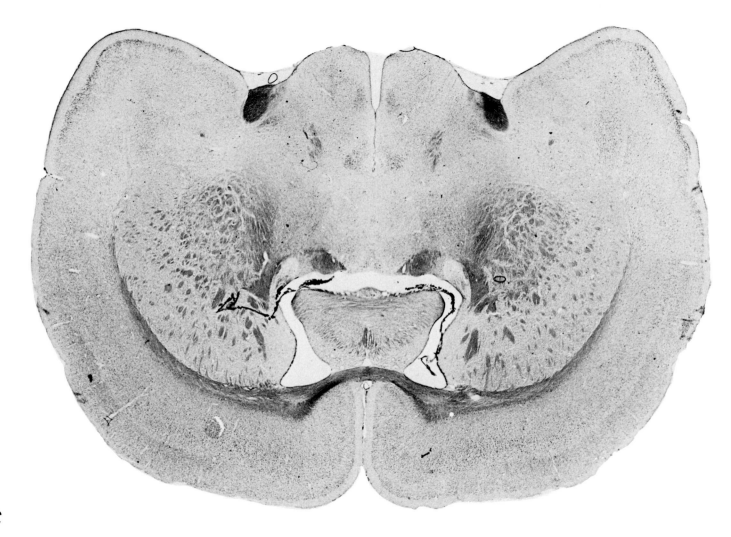

30

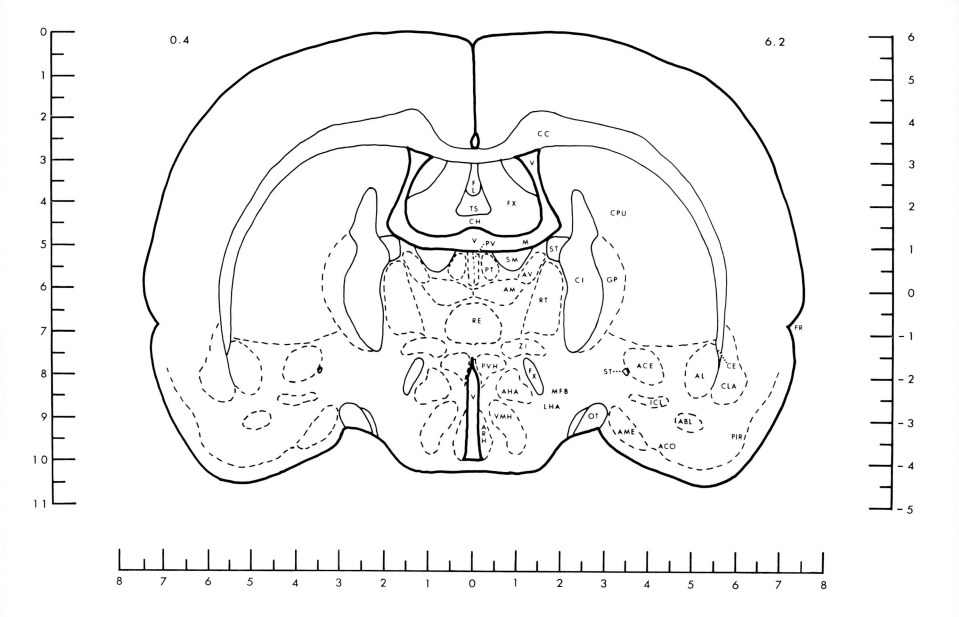

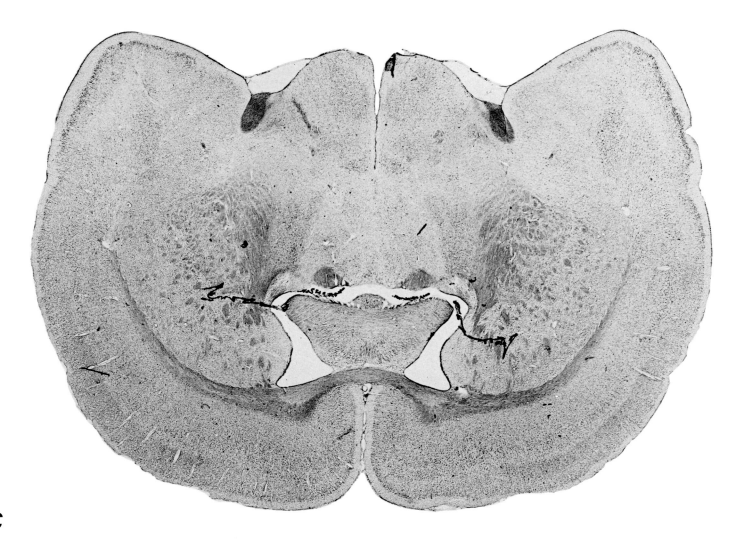

31

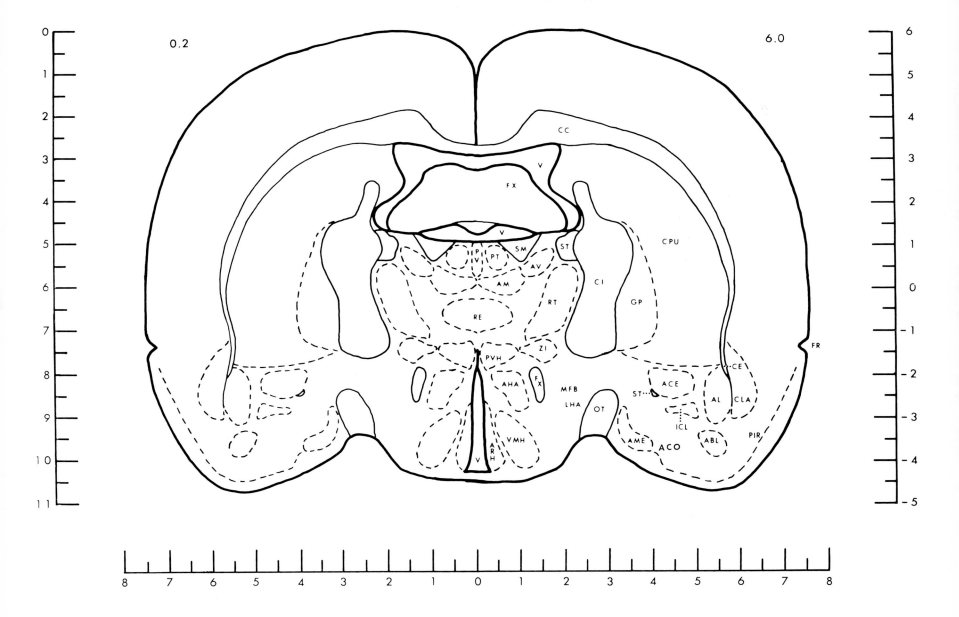

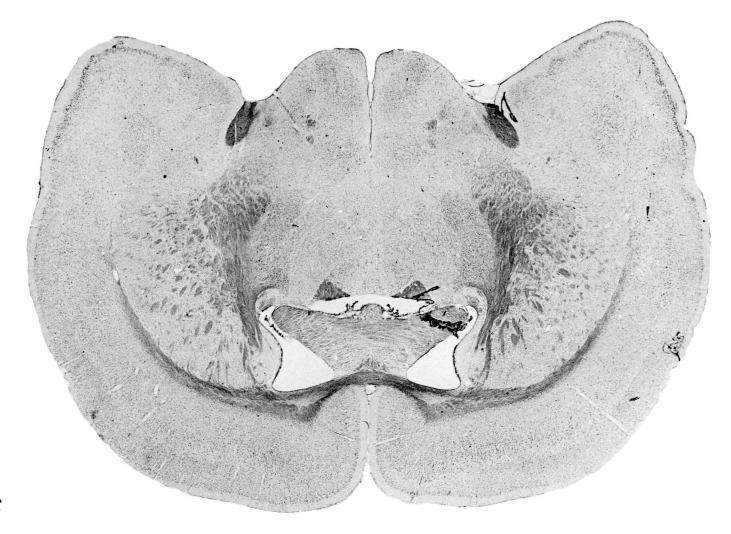

32

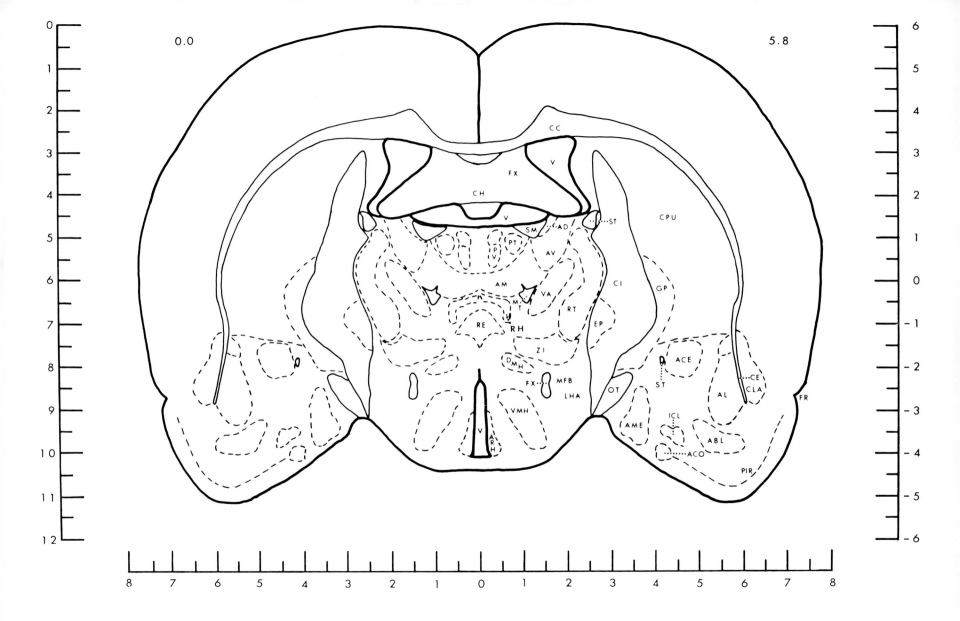

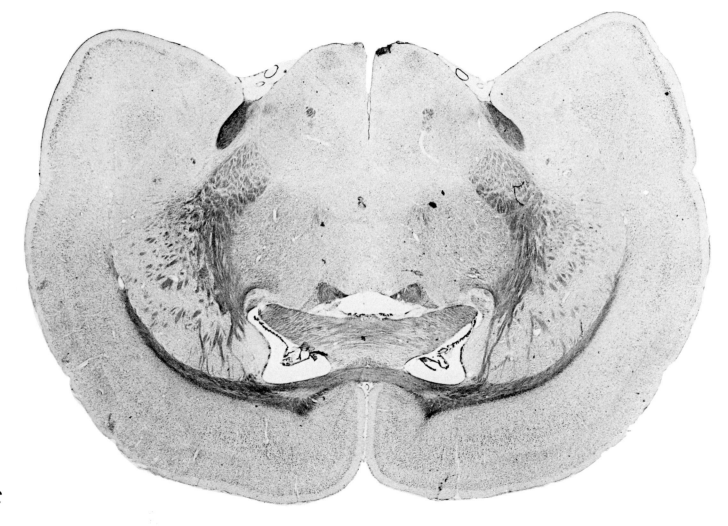

33

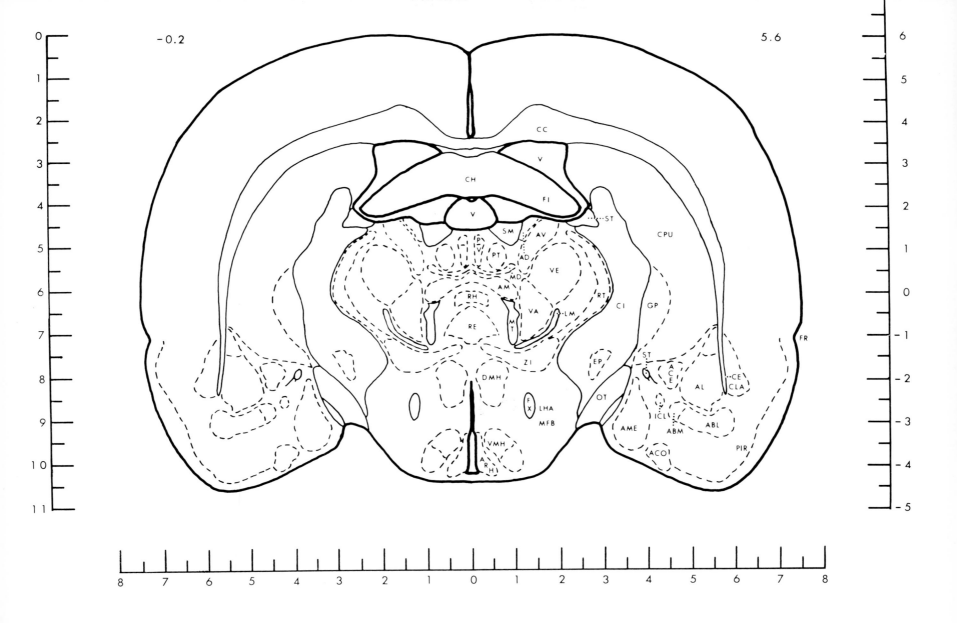

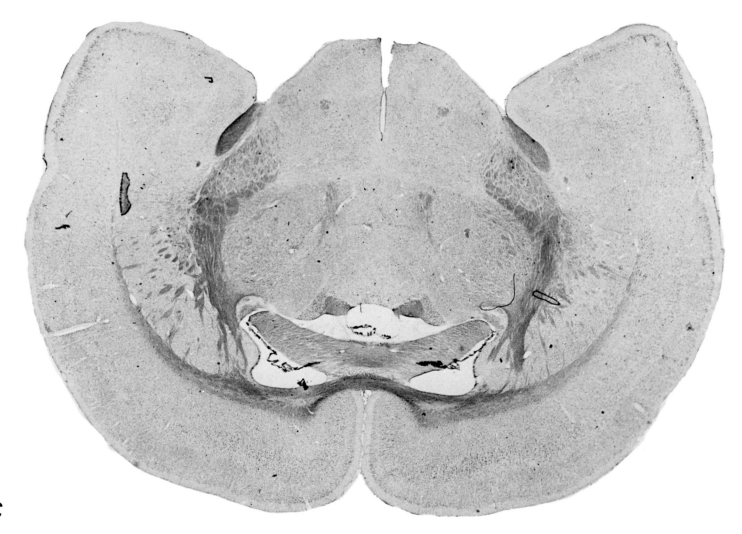

34

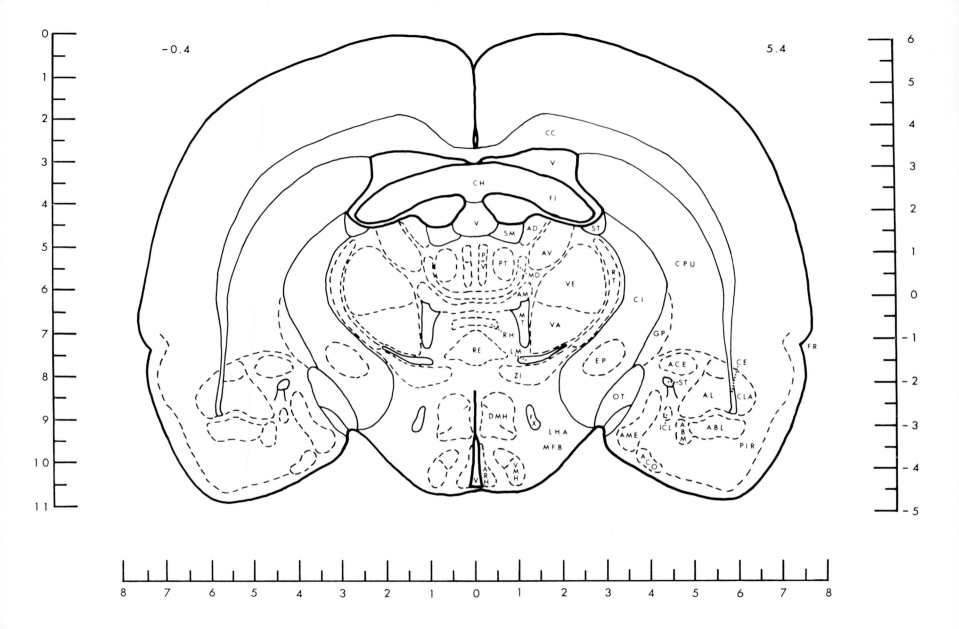

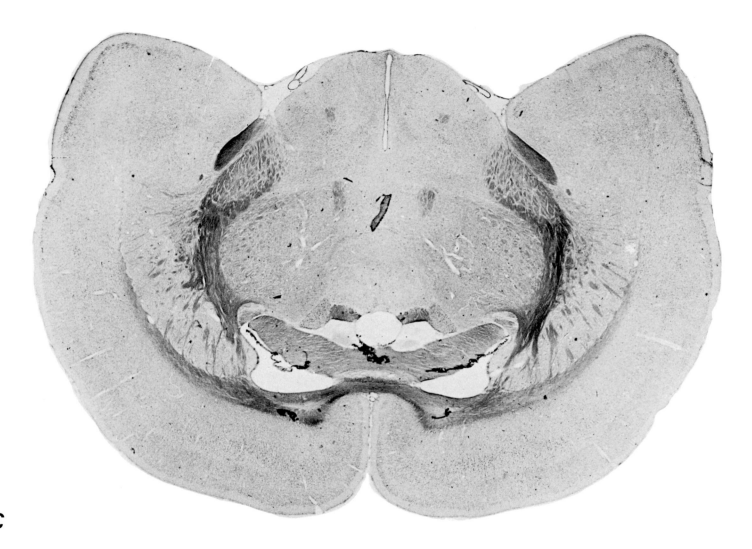

35

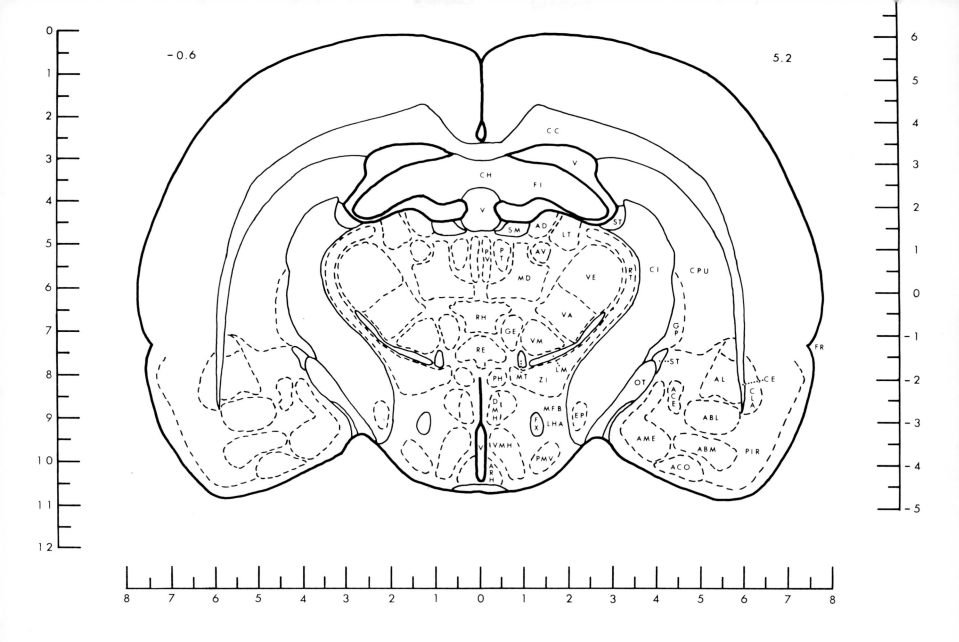

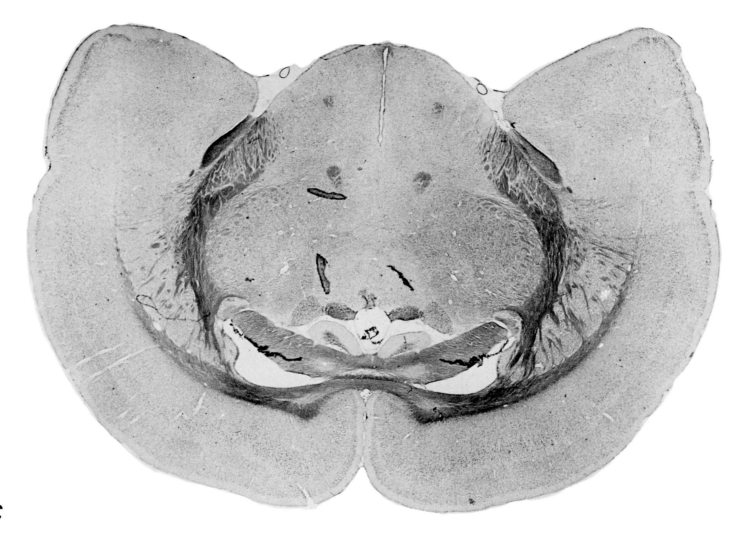

36

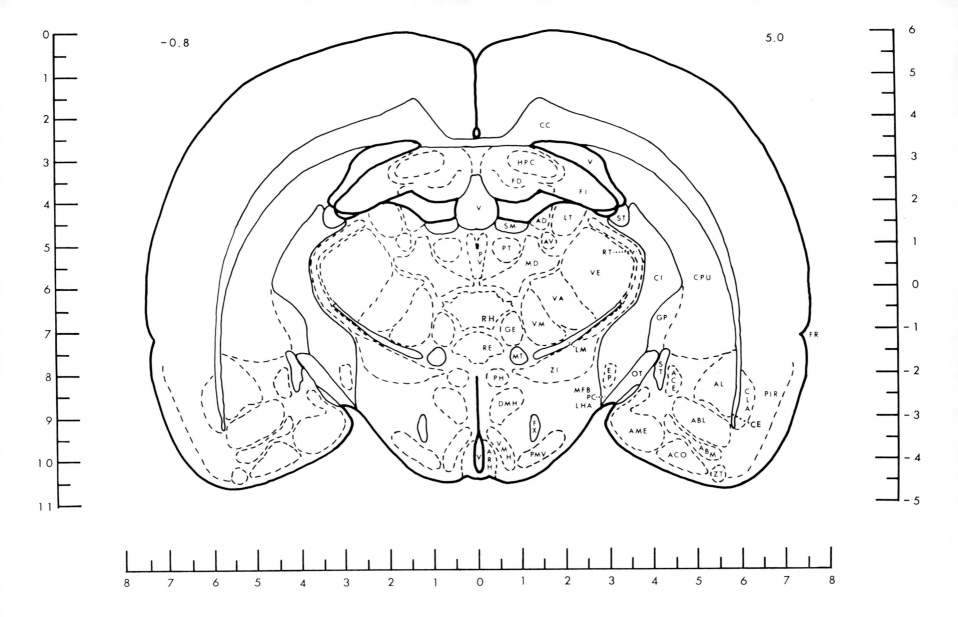

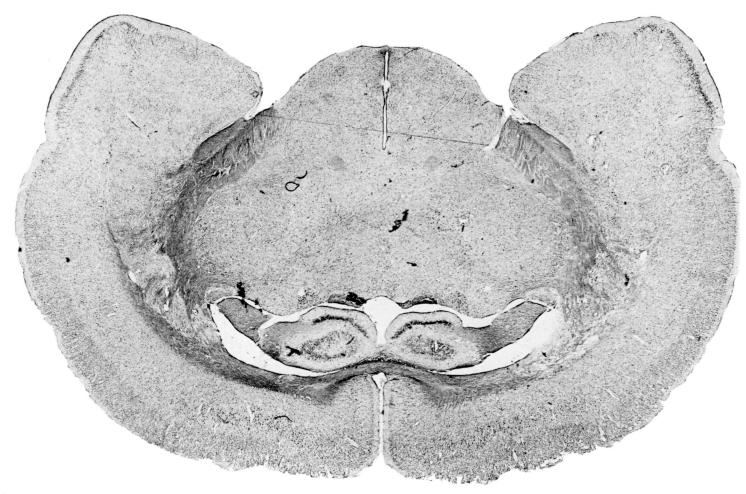

37

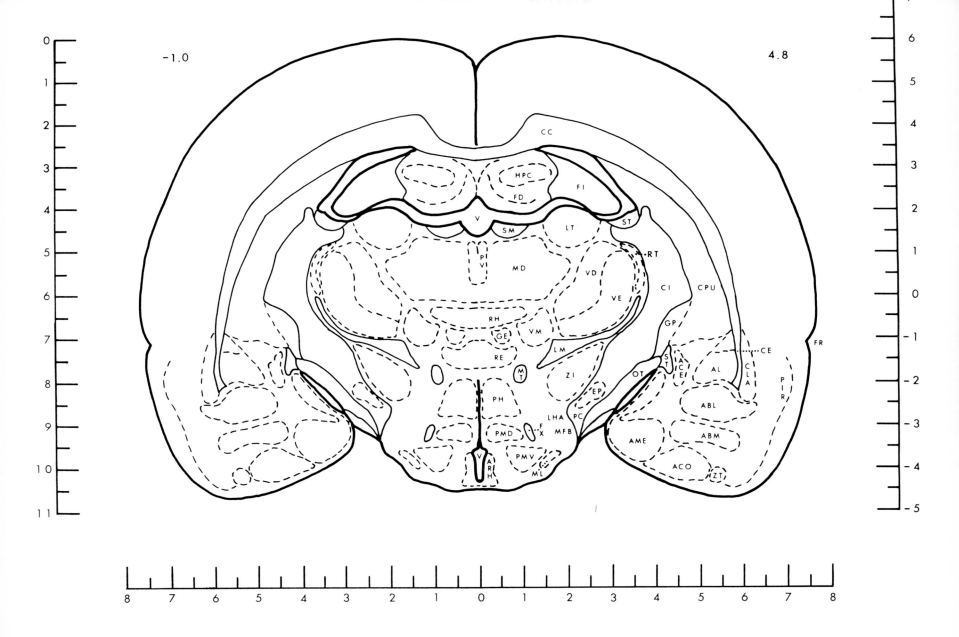

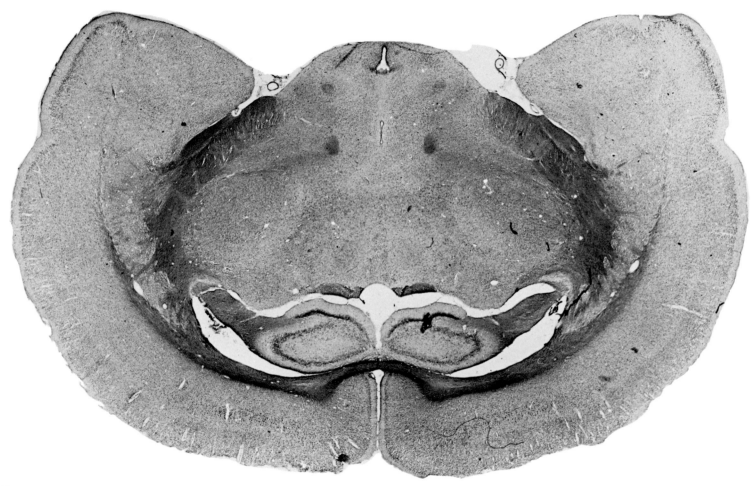

38

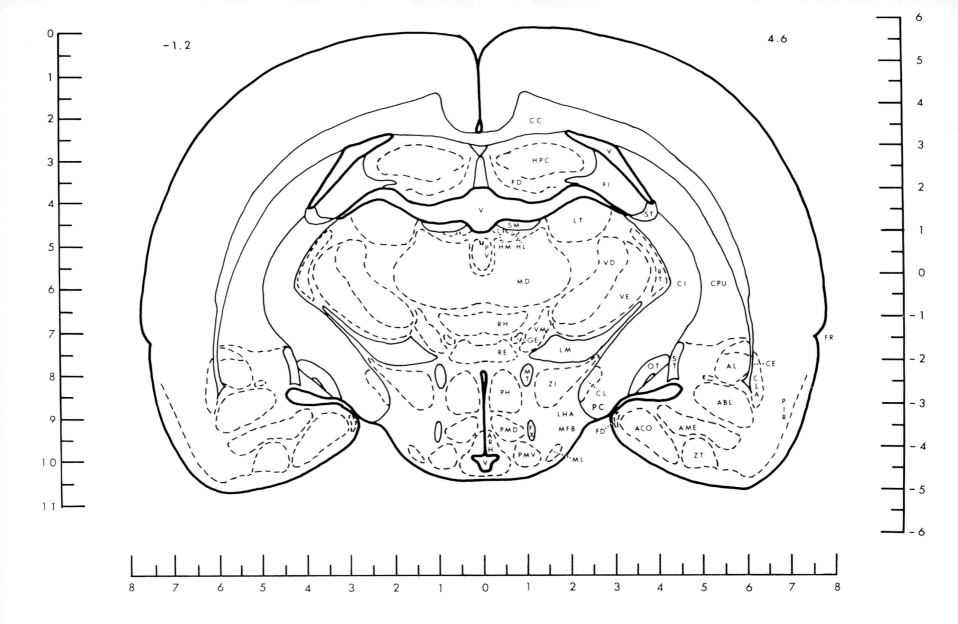

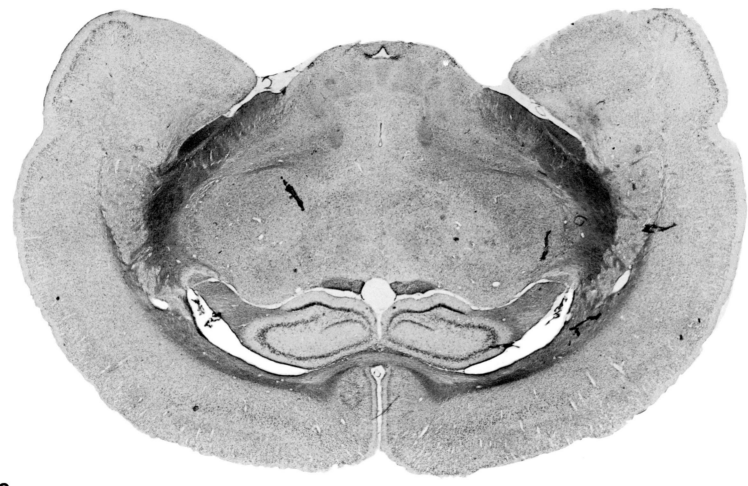

39

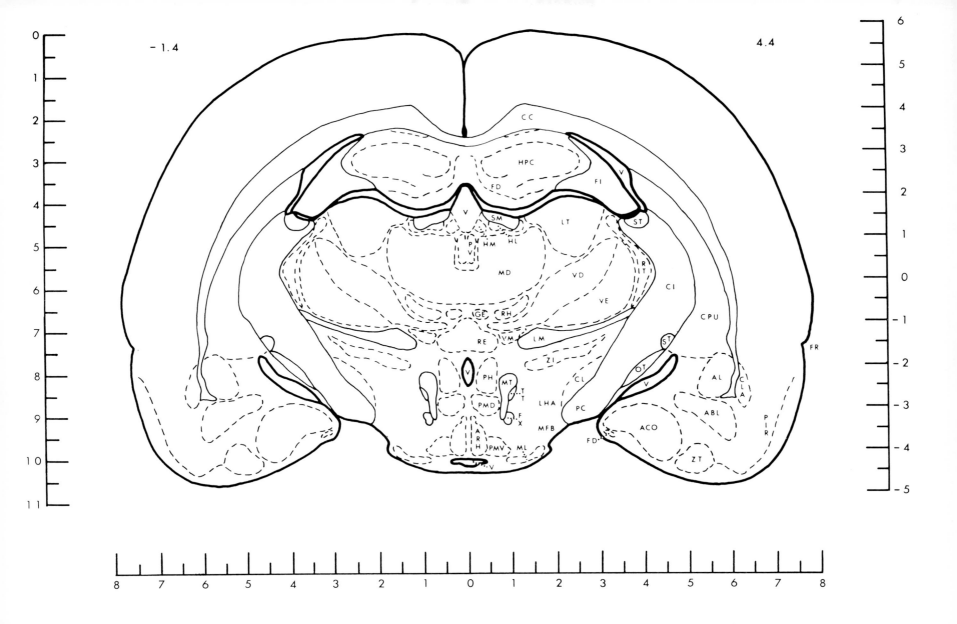

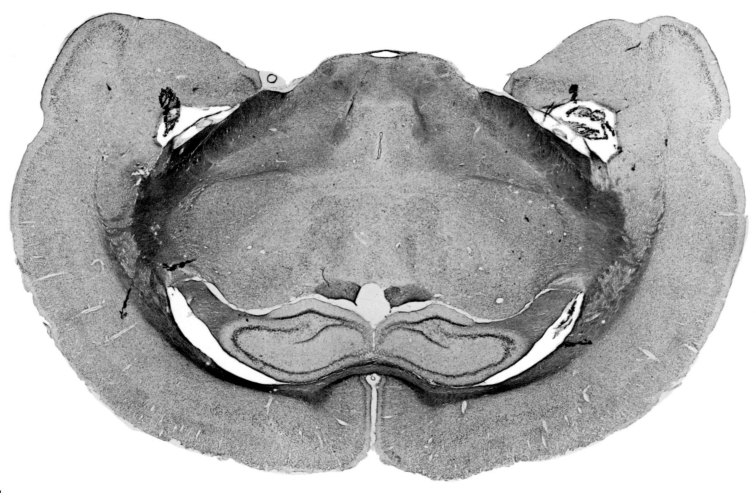

40

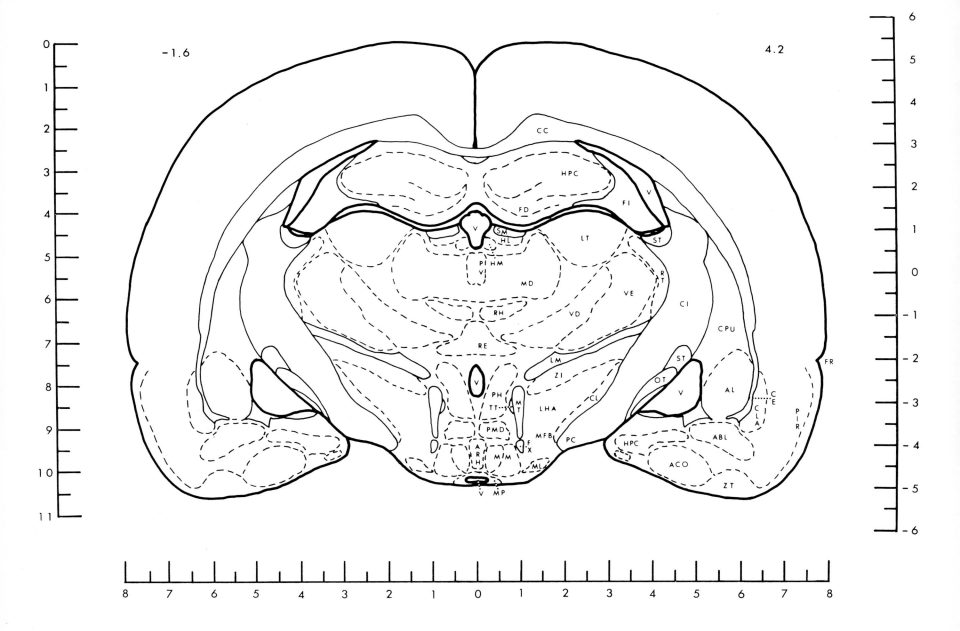

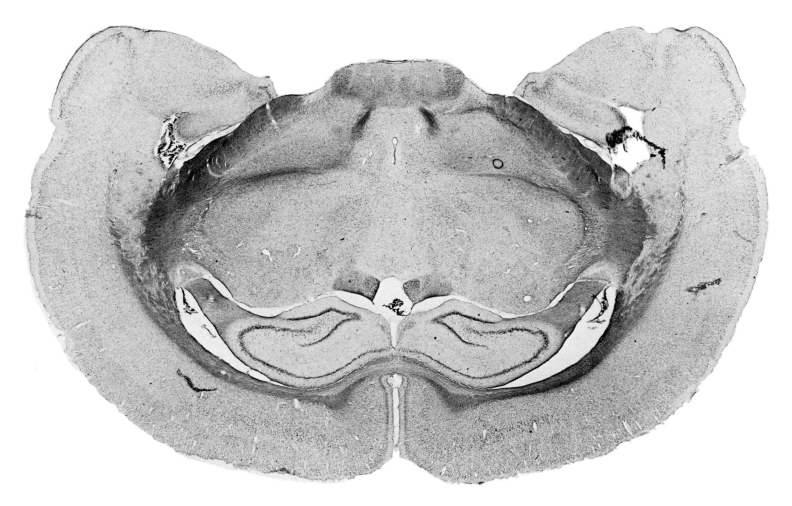

41

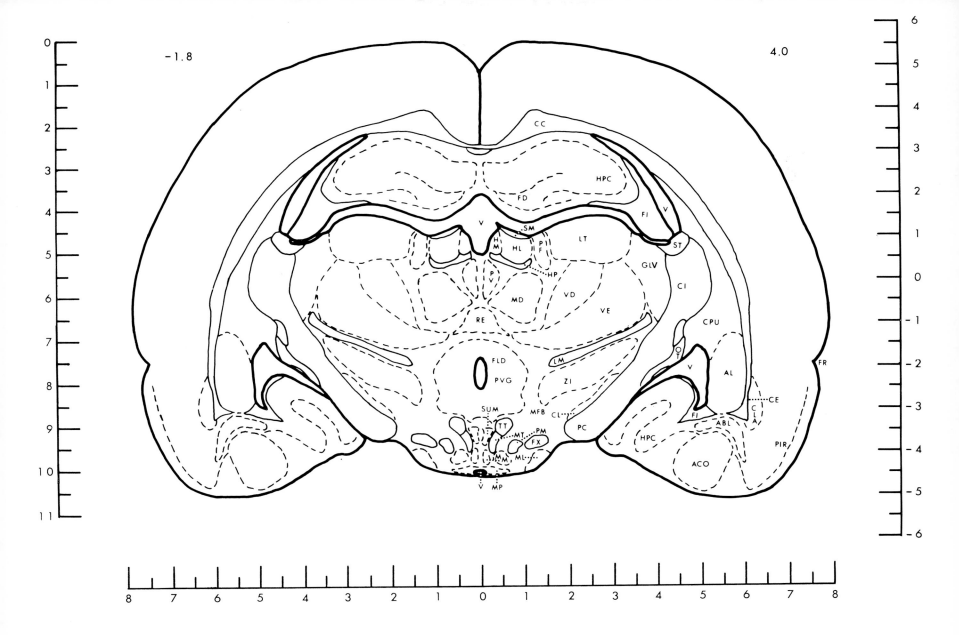

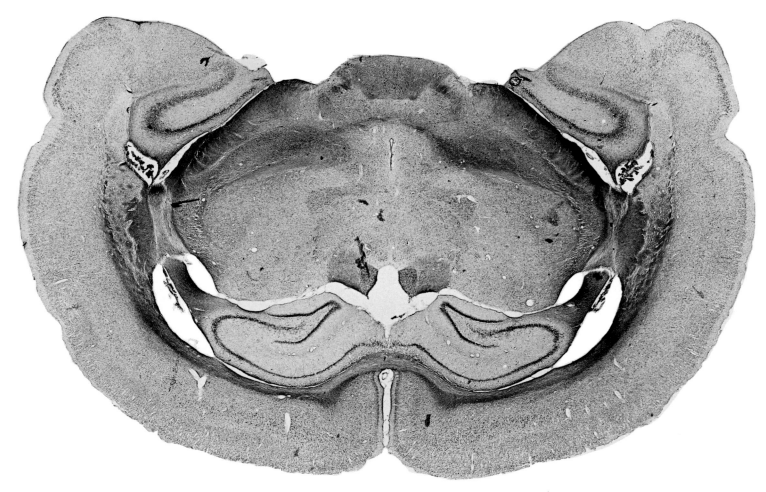

42

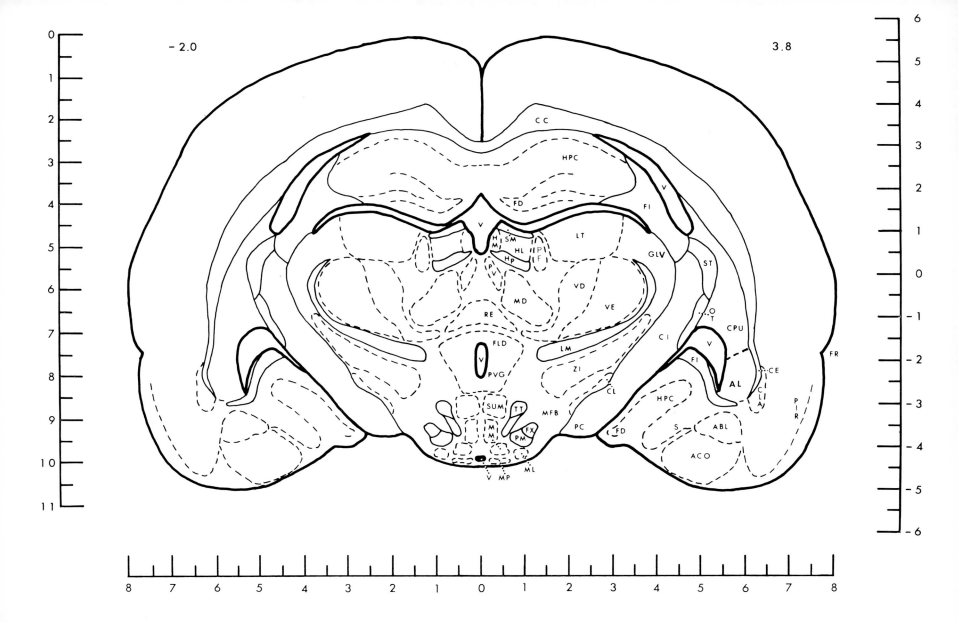

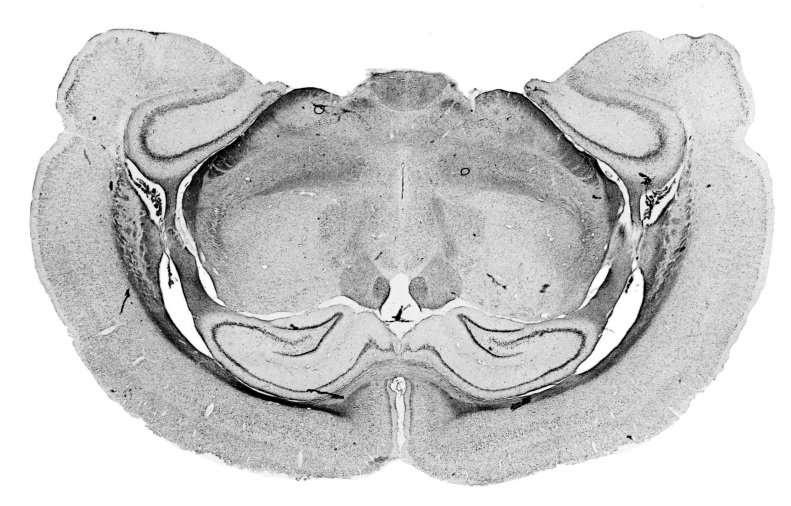

43

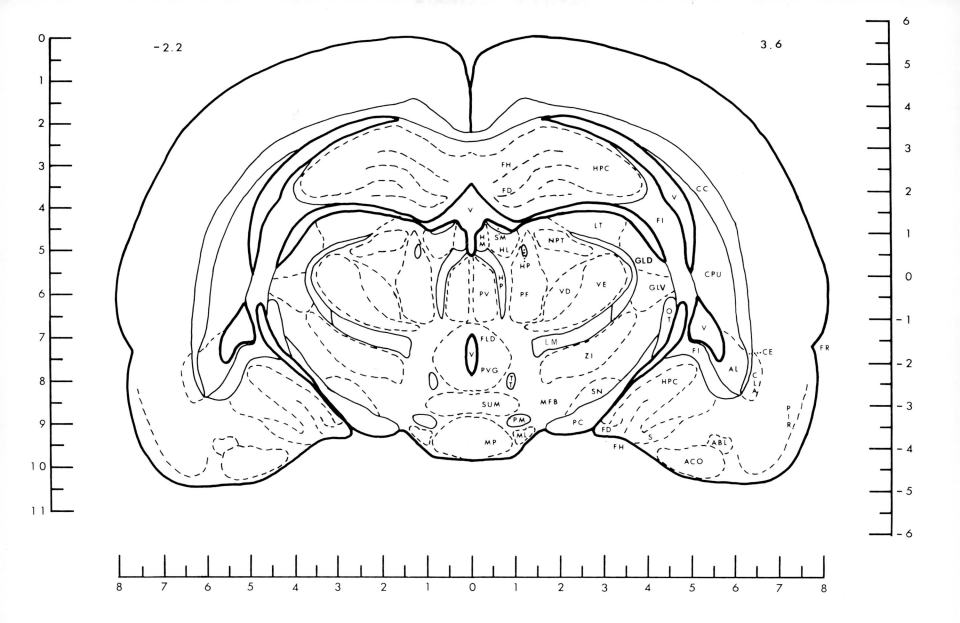

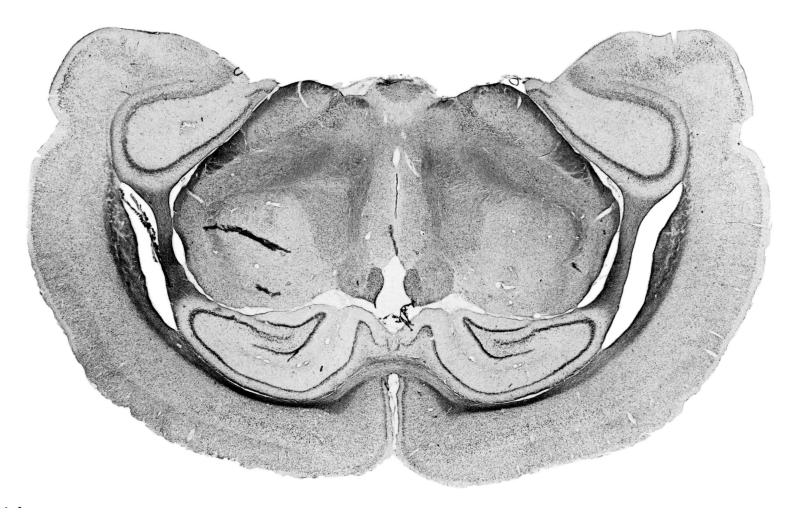

44

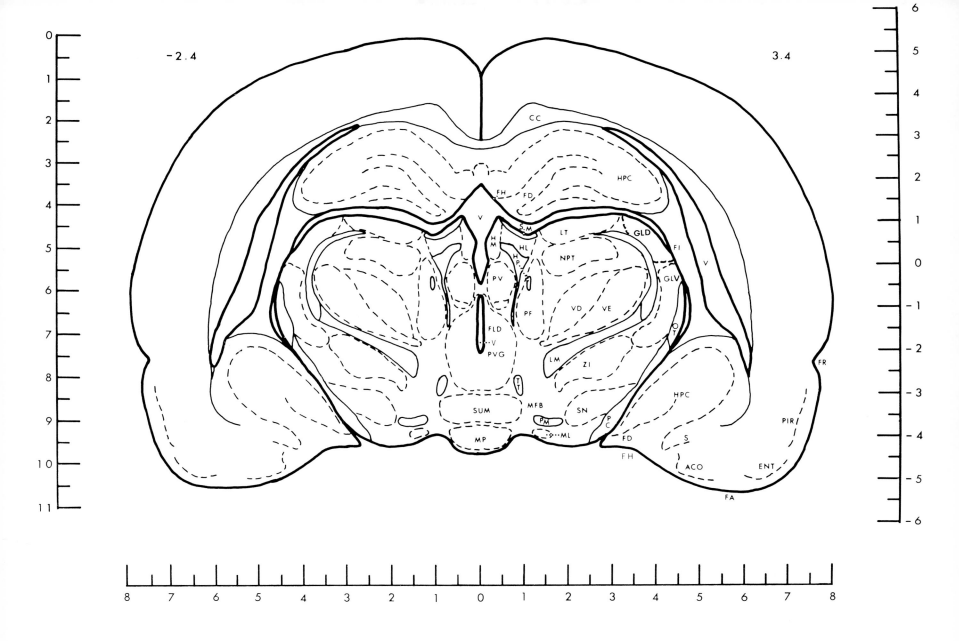

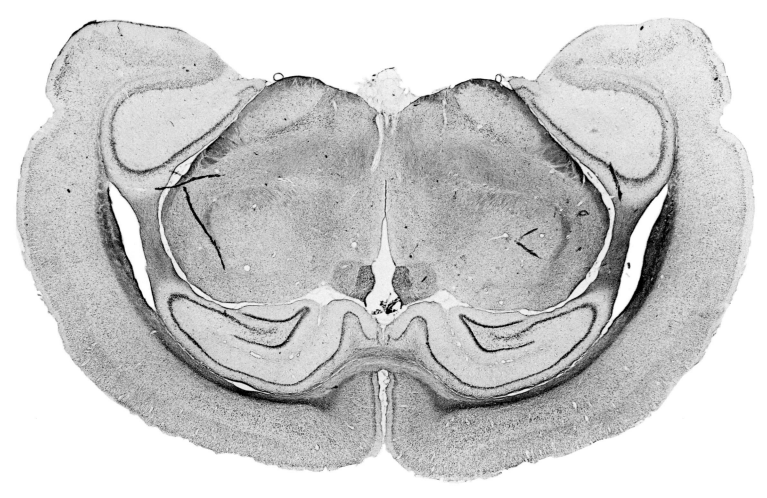

45

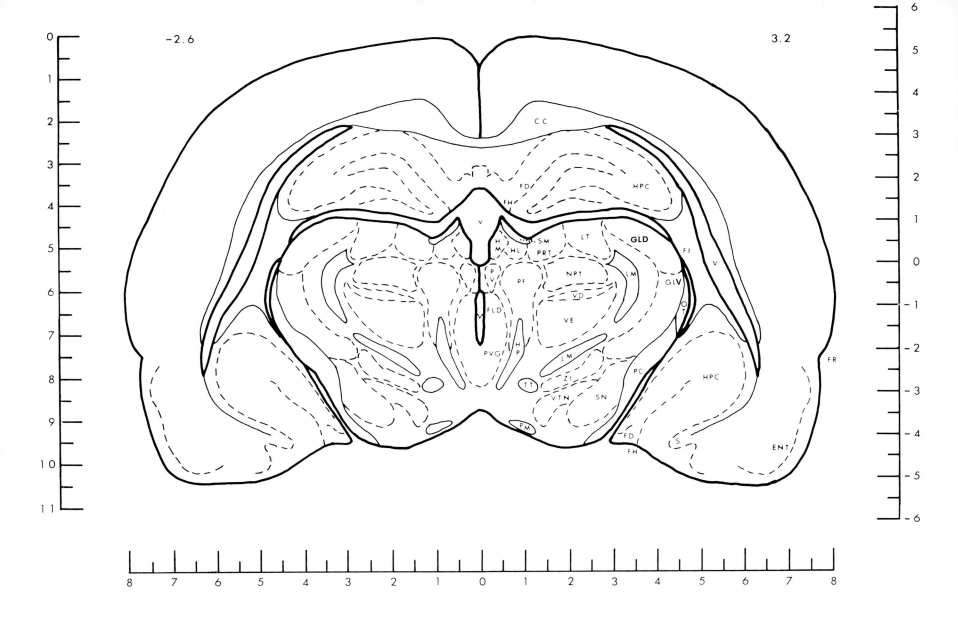

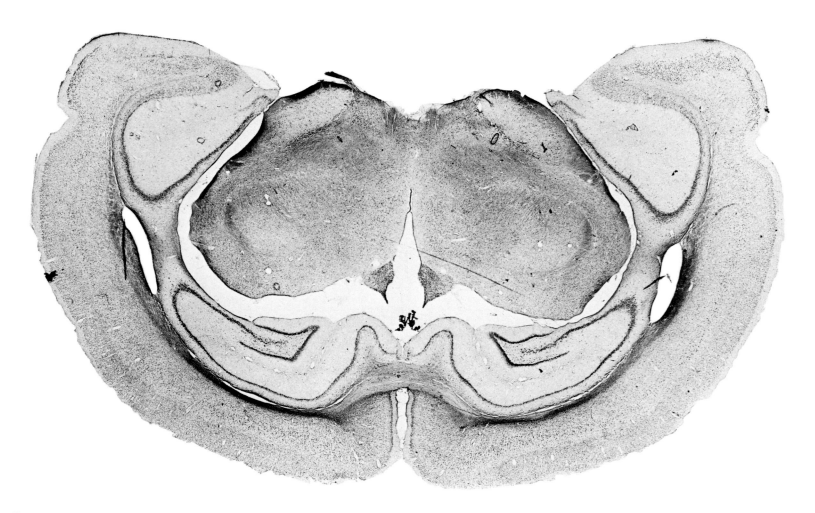

46

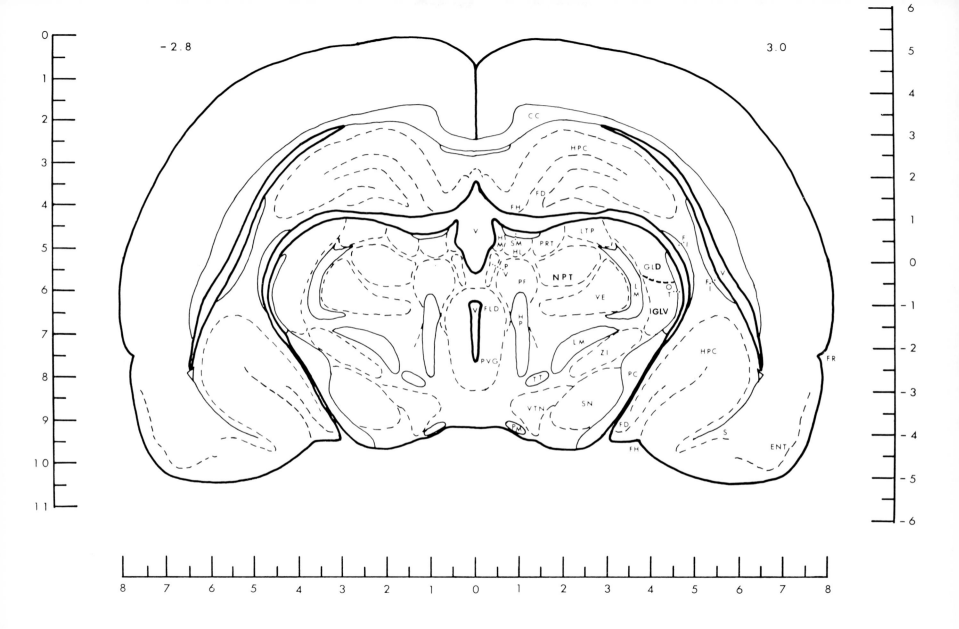

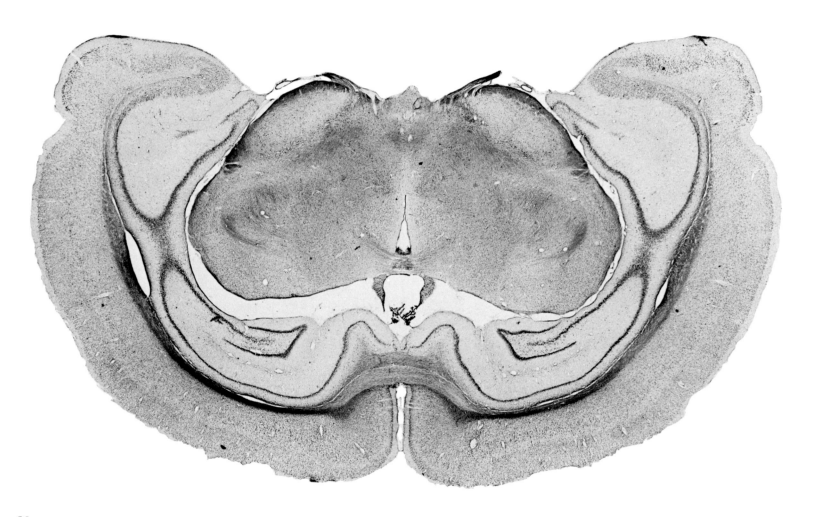

47

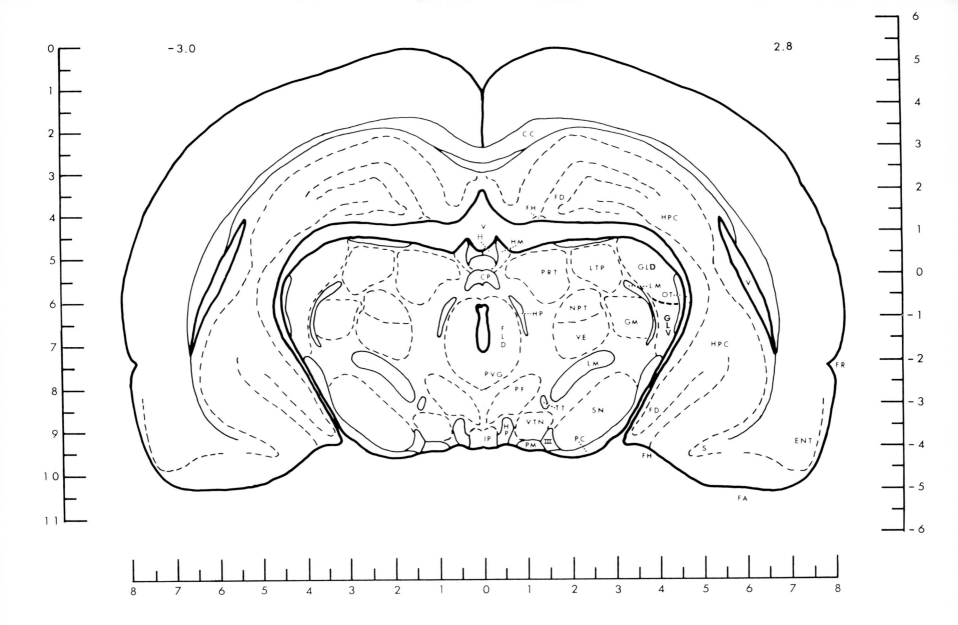

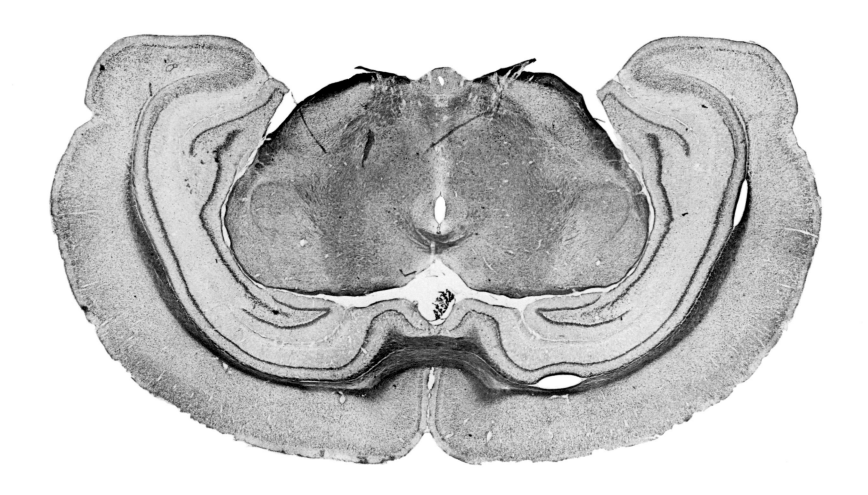

48

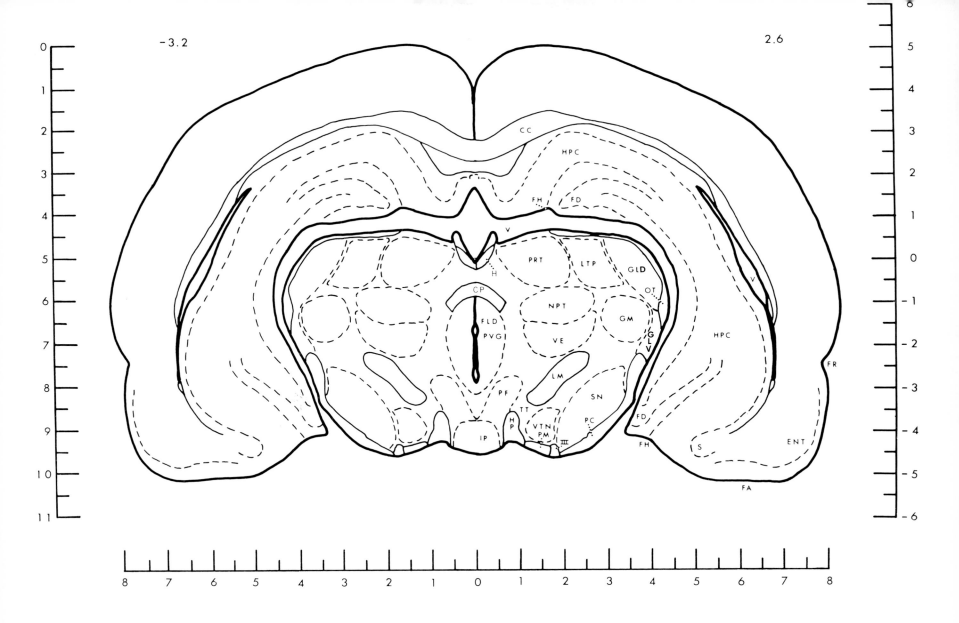

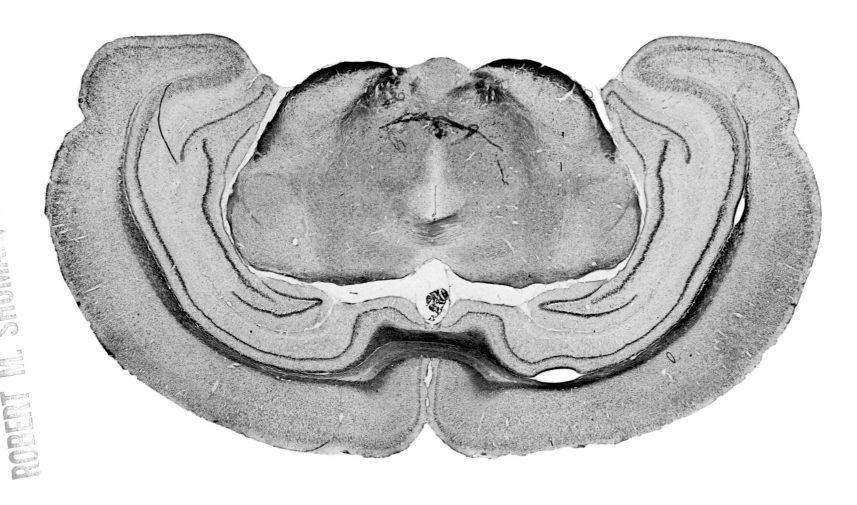

49

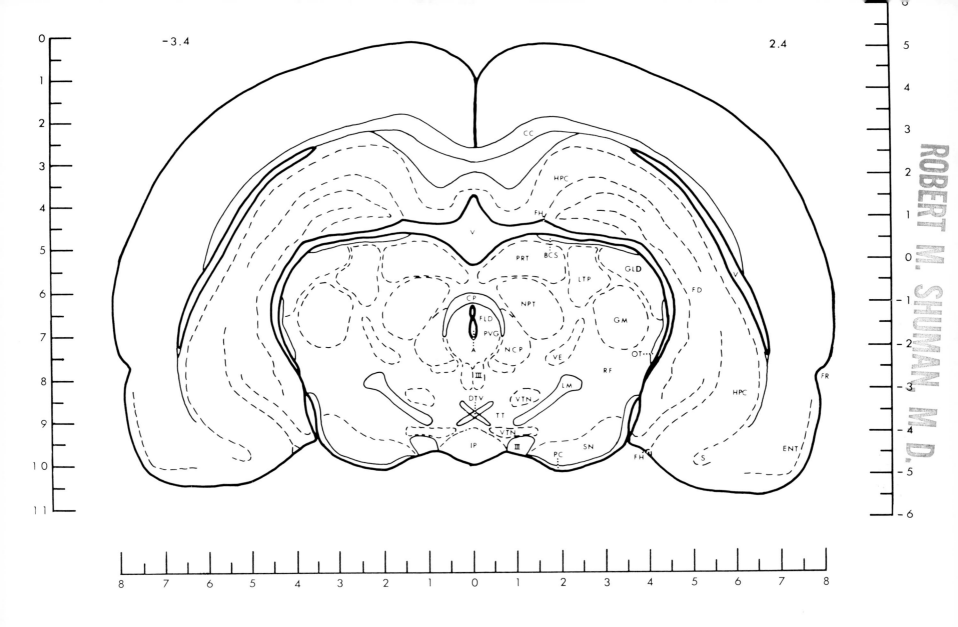

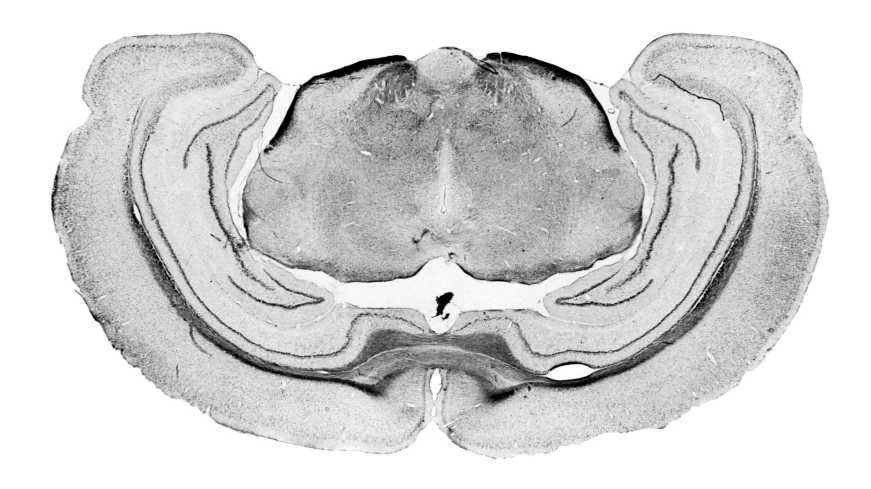

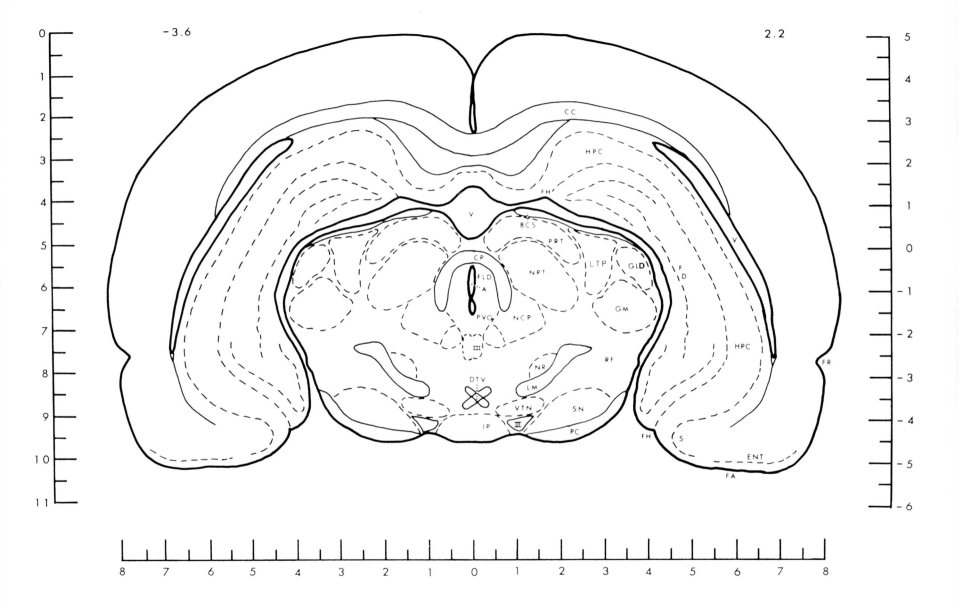

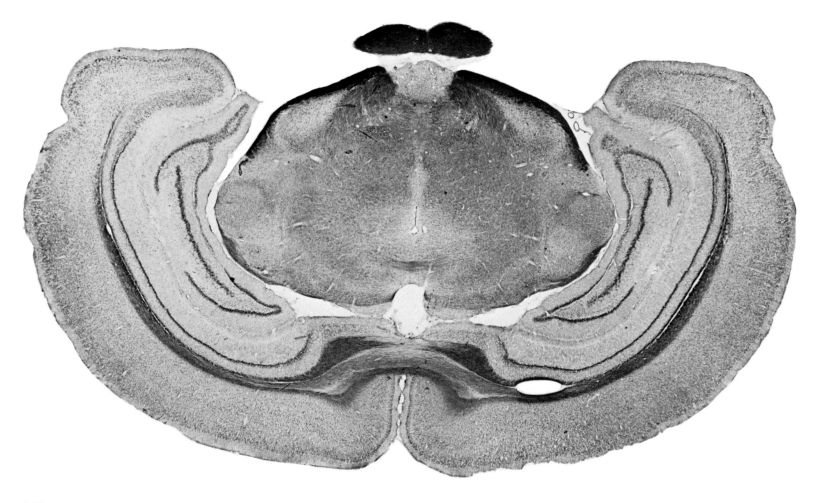

51

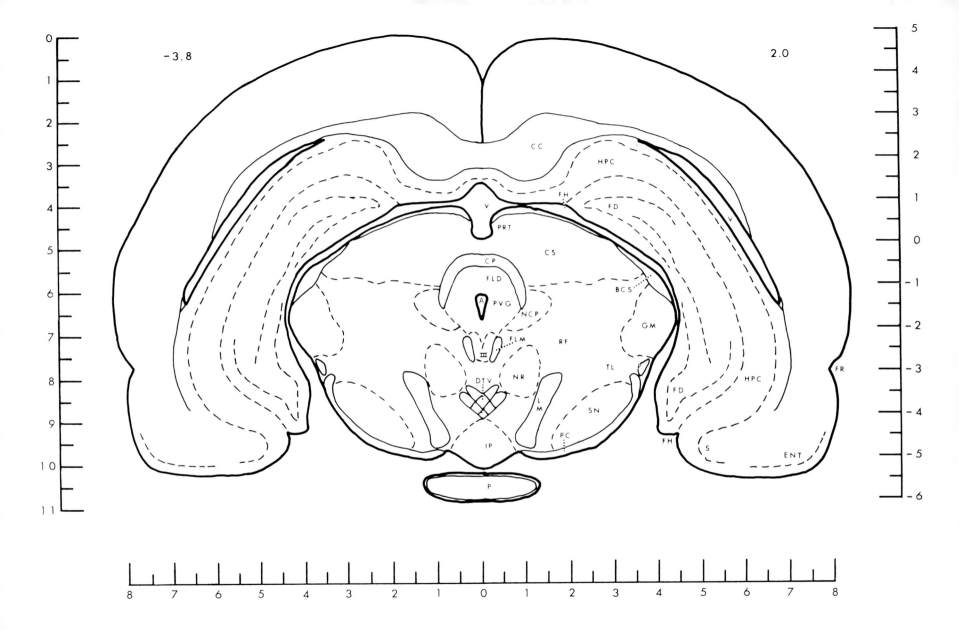

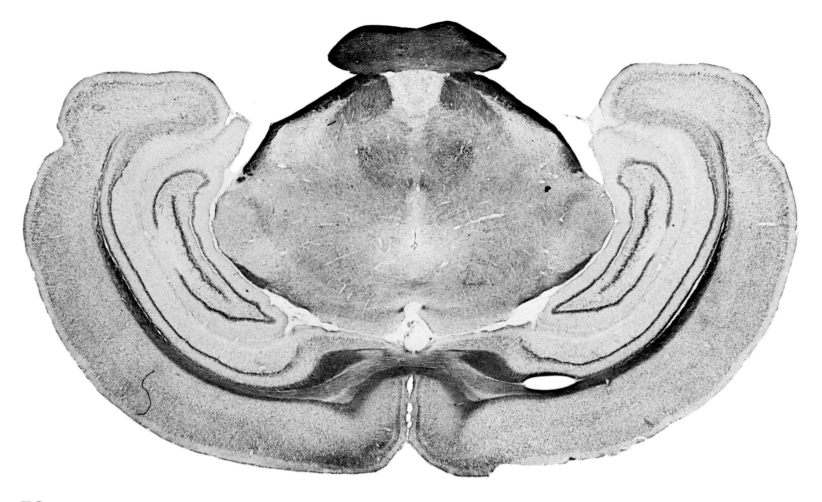

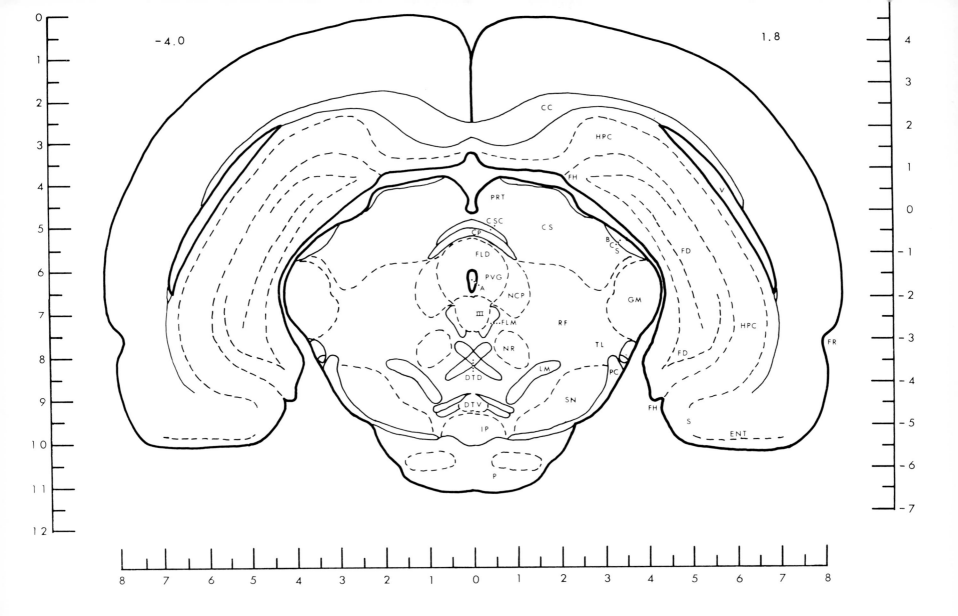

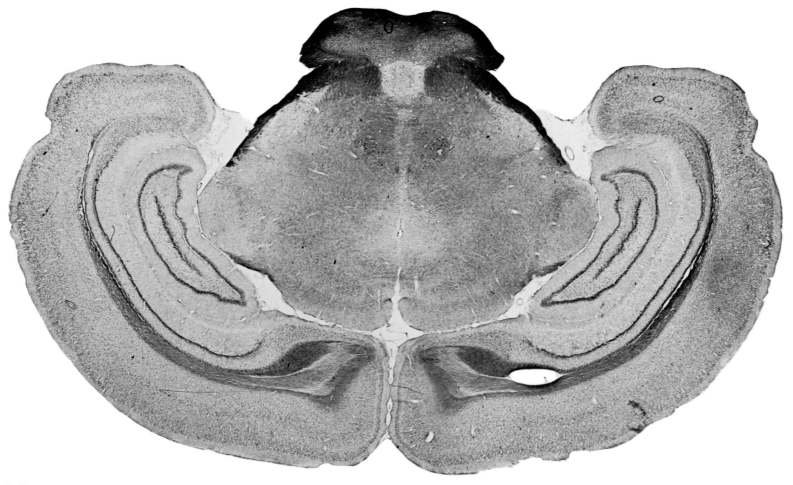

53

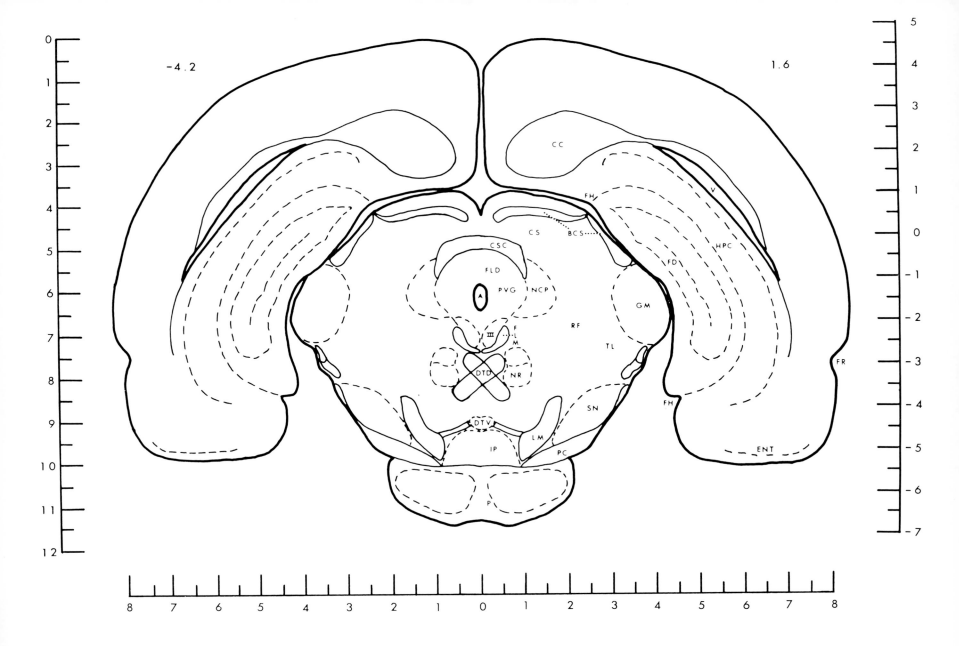

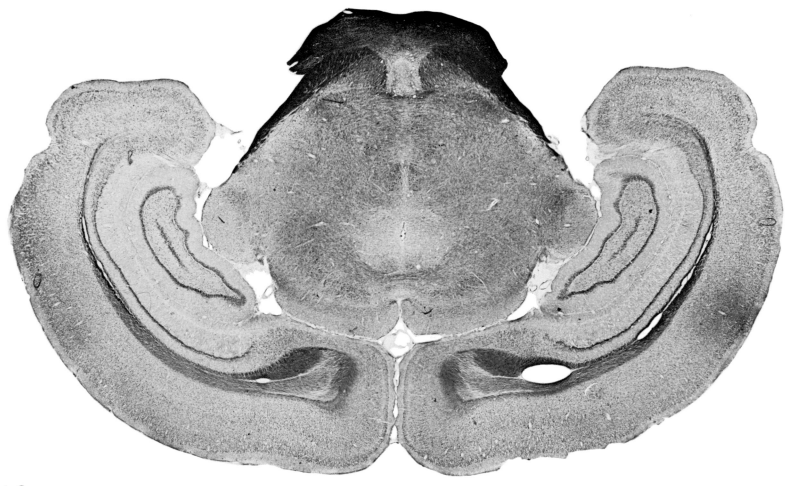

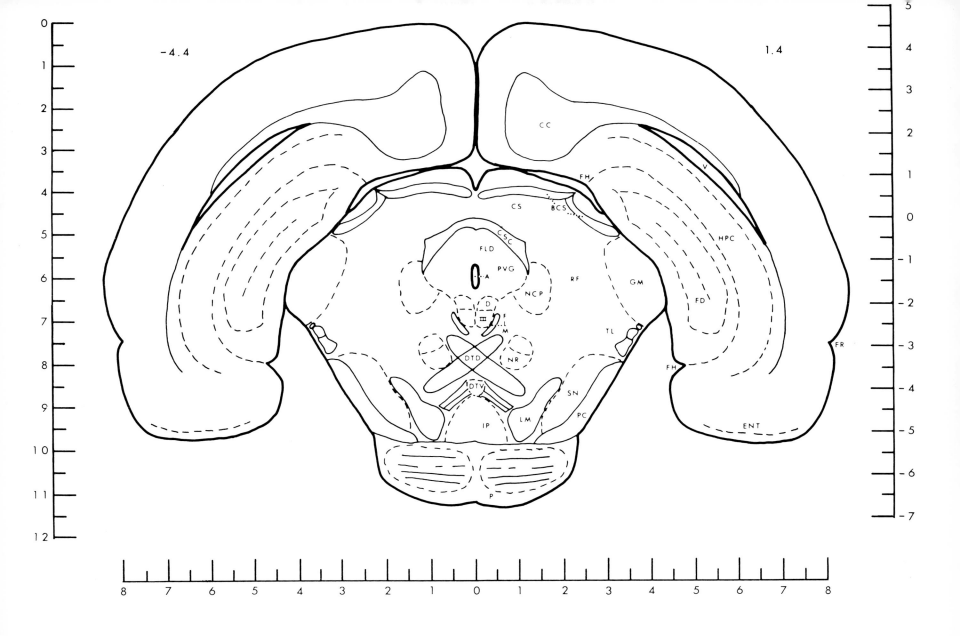

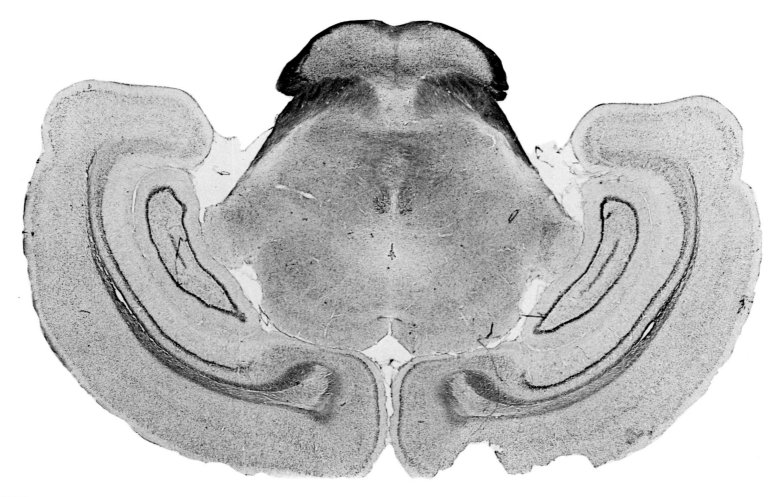

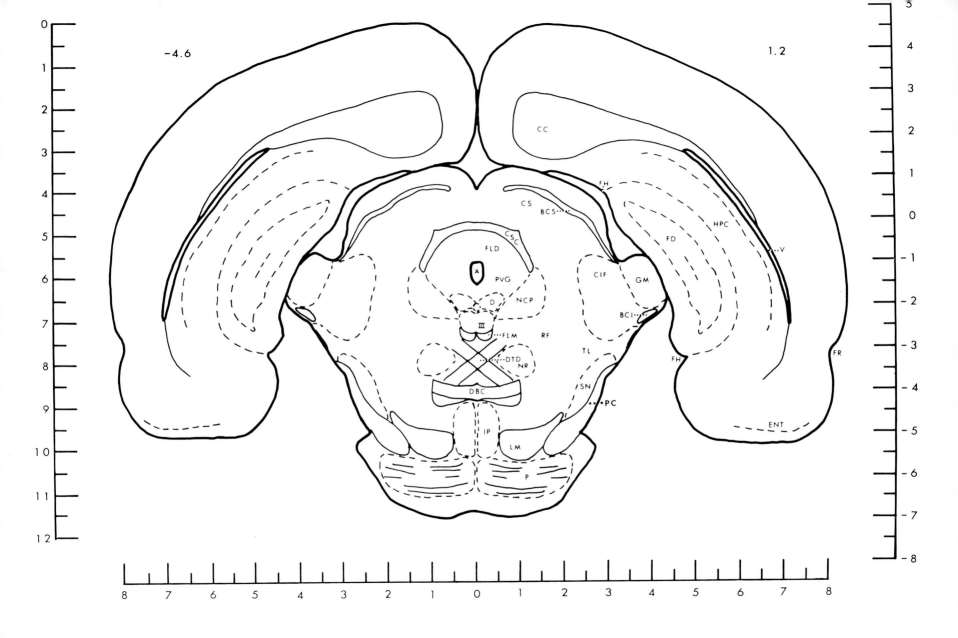

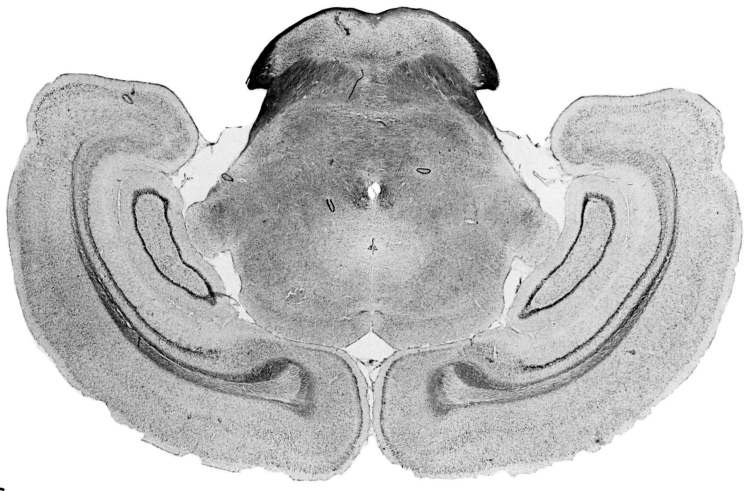

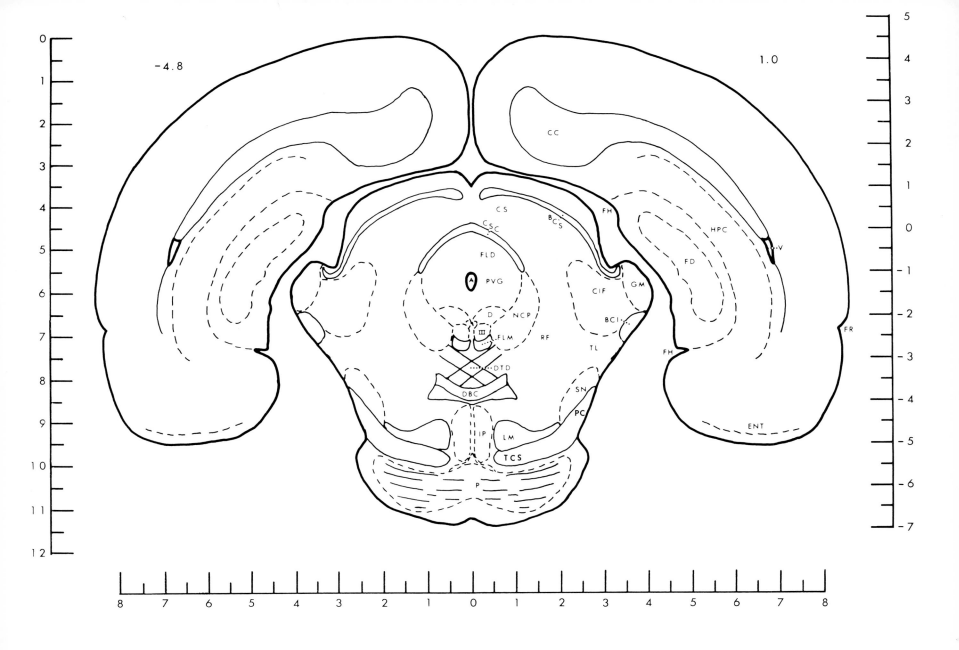

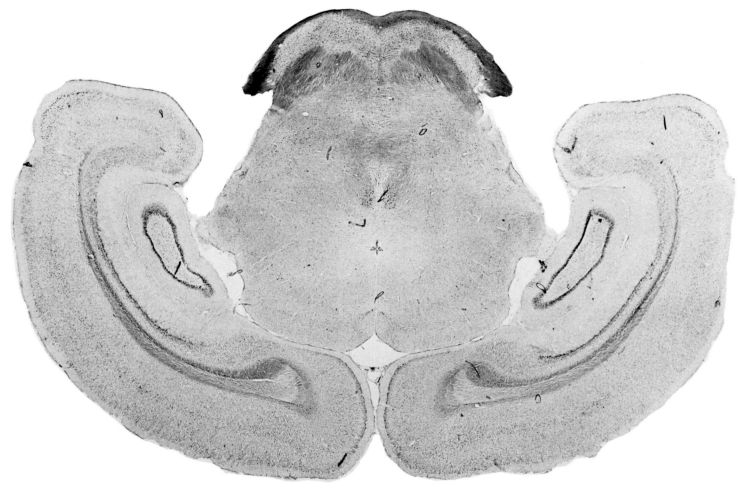

57

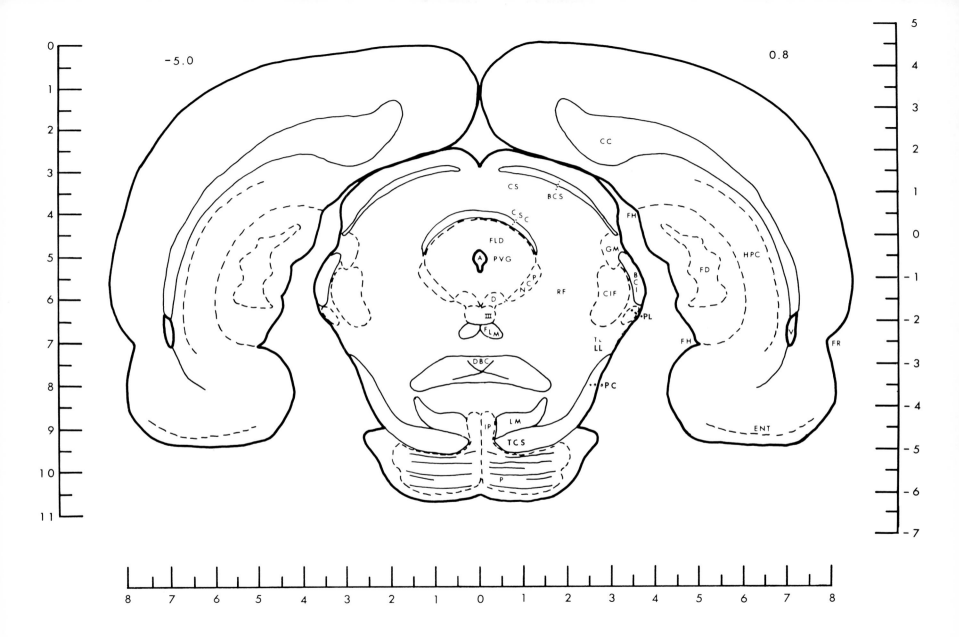

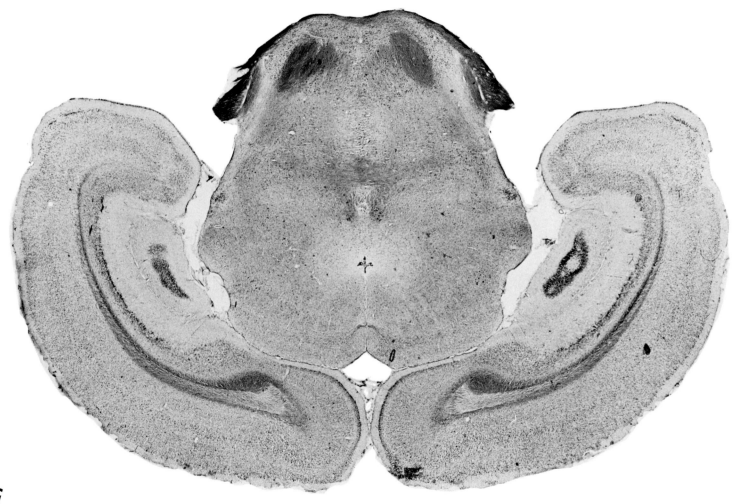

58

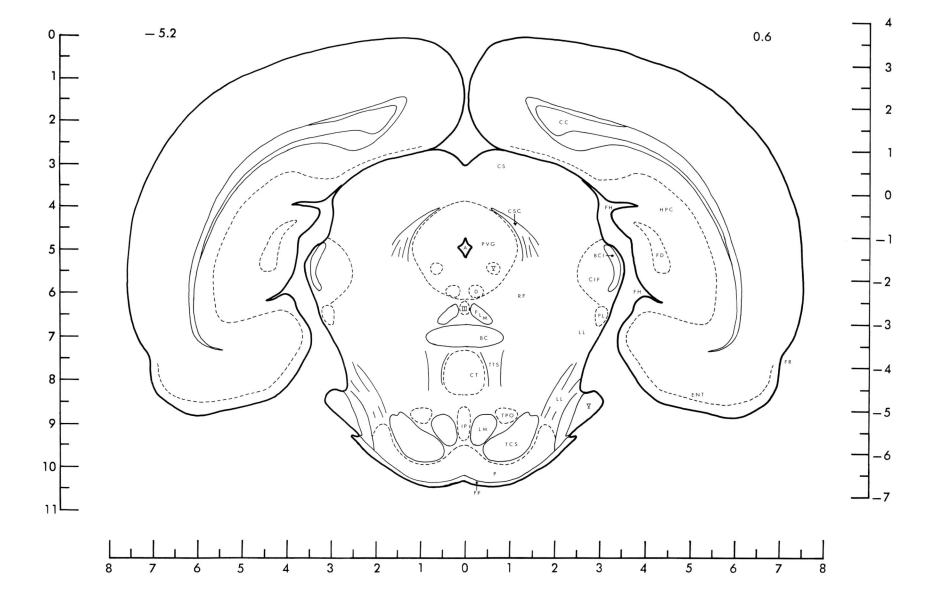

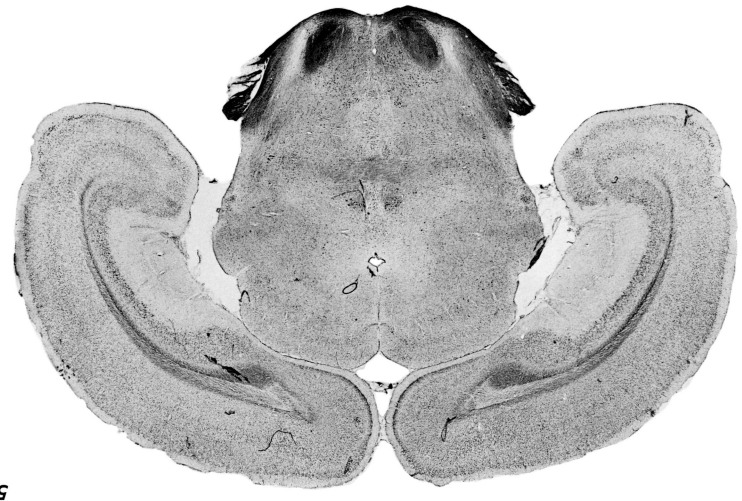

59

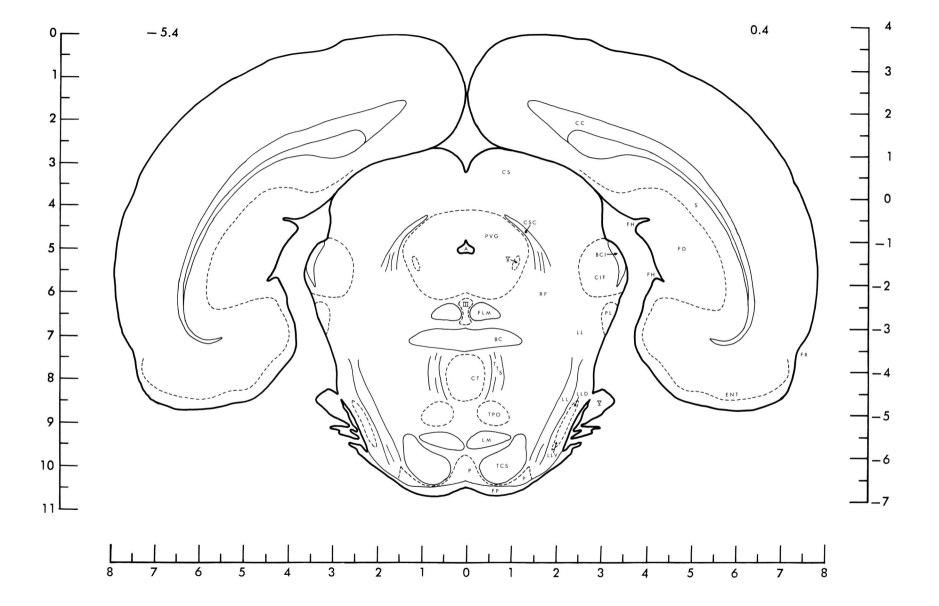

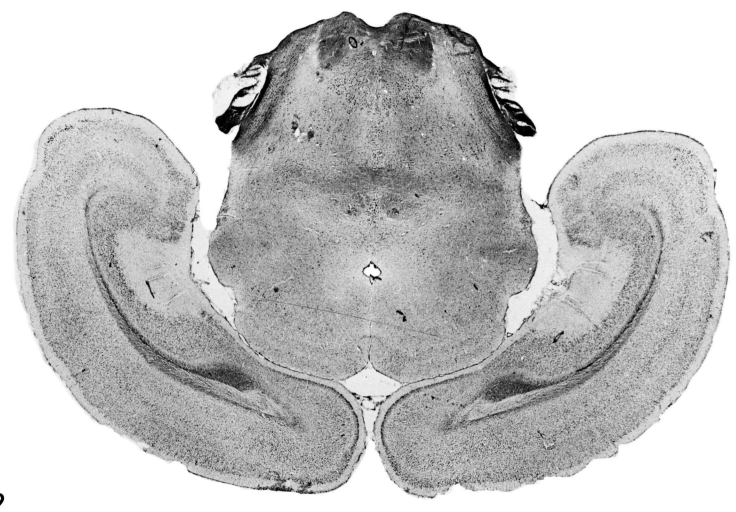

09

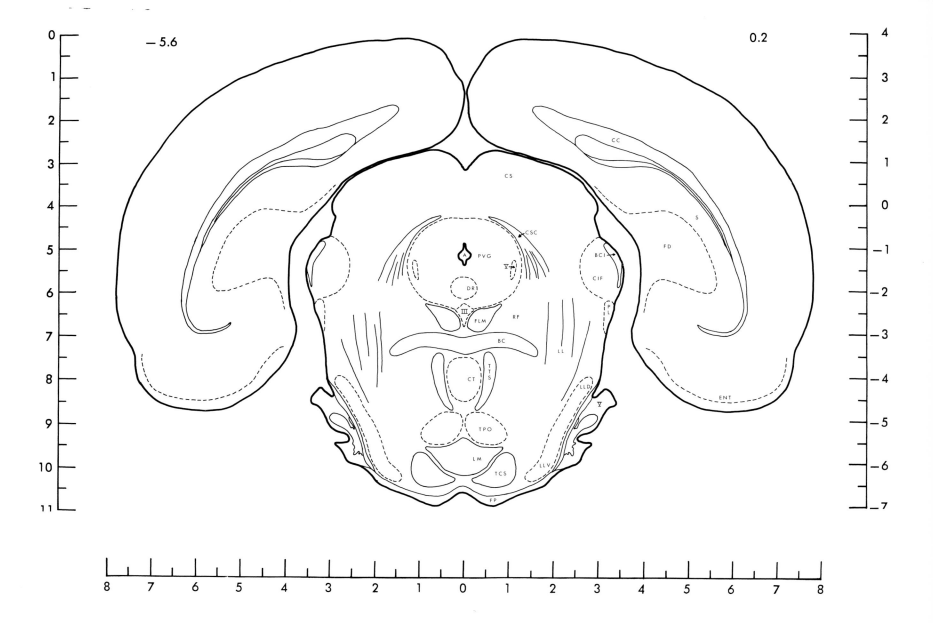

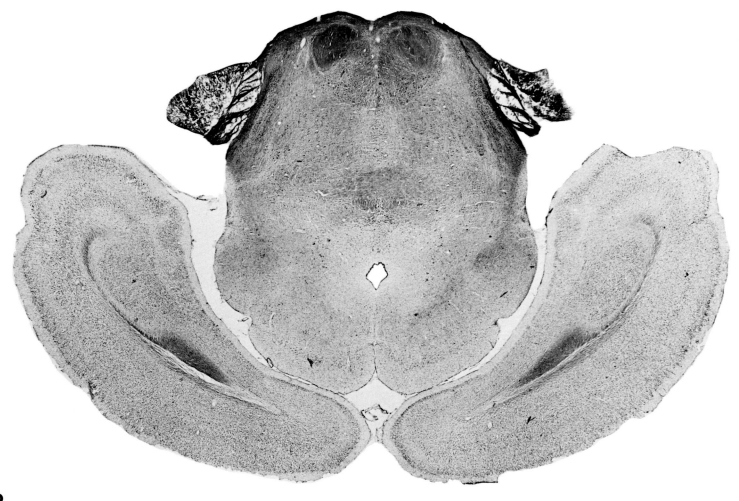

61

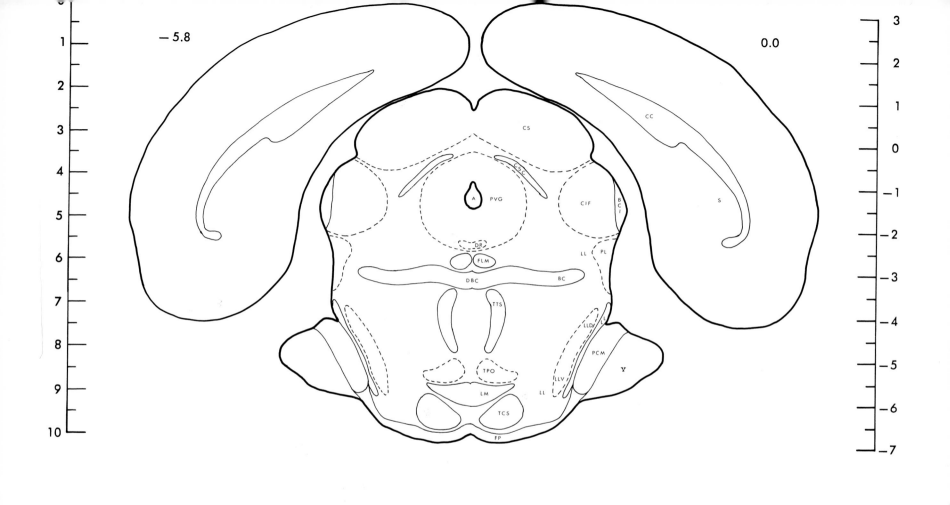

−5.8

0.0

CS

CC

CSC

A PVG

CIF

BCI

S

DR

LL PL

FLM

DBC BC

TTS

LLD L

PCM

Y

TPO

LLV

LM

LL

TCS

FP

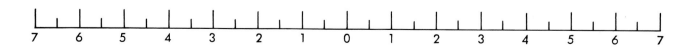

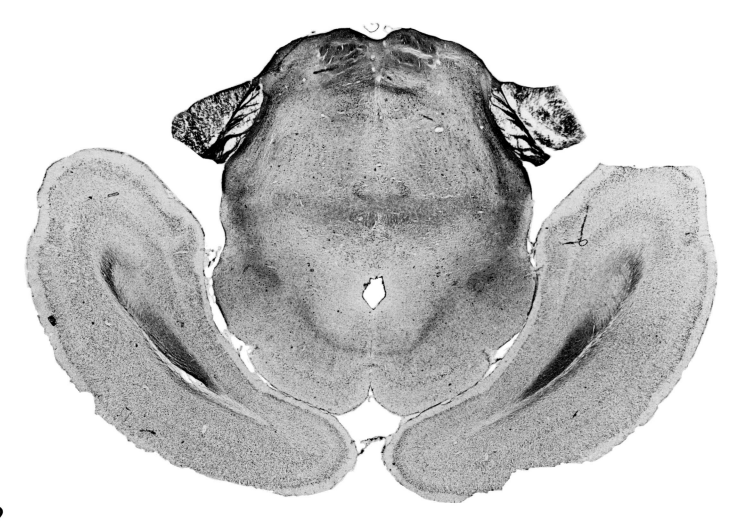

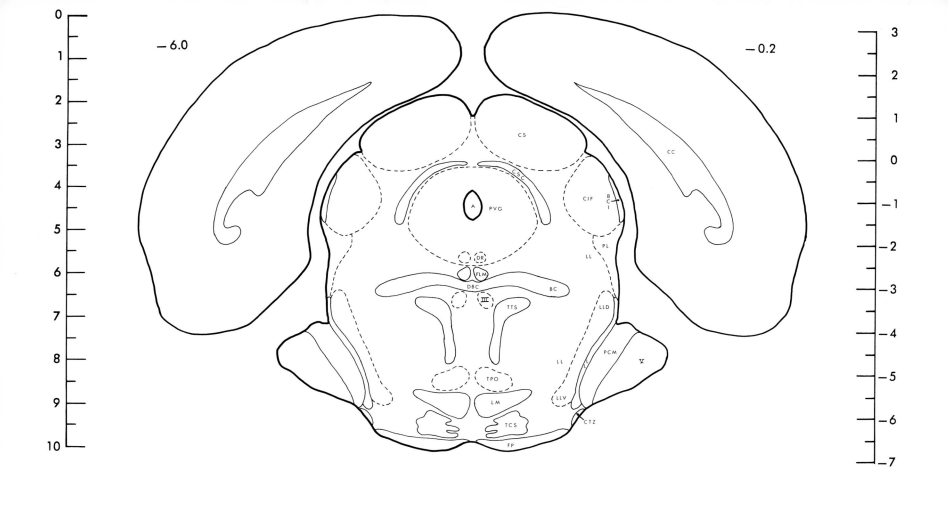

−6.0

−0.2

CS

CC

CSC

CIF

A PVG

B
C
I

PL

LL

DR

FLM

DBC BC

III

TTS

LLD

LL

PCM

V

TPO

LLV

LM

CTZ

TCS

FP

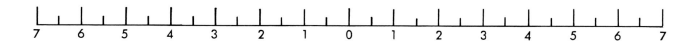

7 6 5 4 3 2 1 0 1 2 3 4 5 6 7

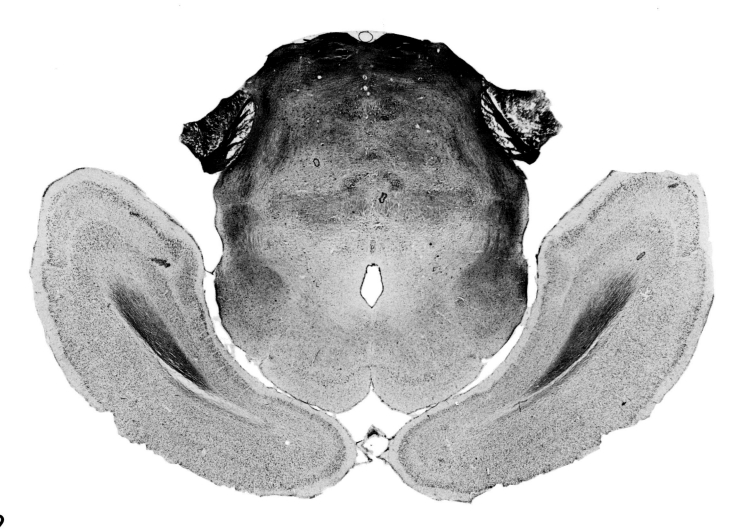

63

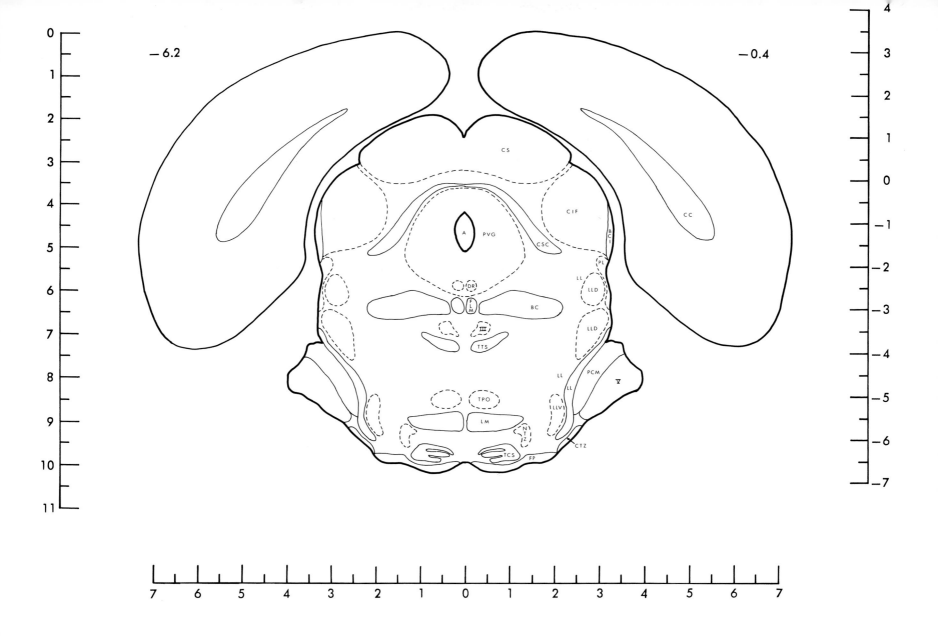

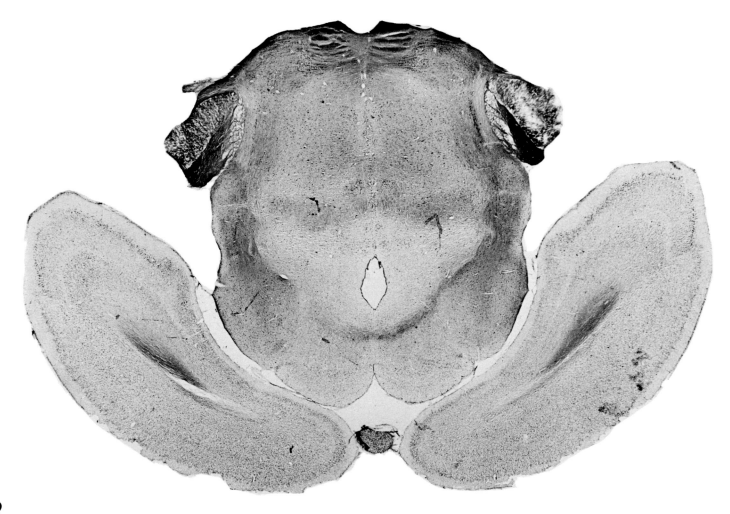

64

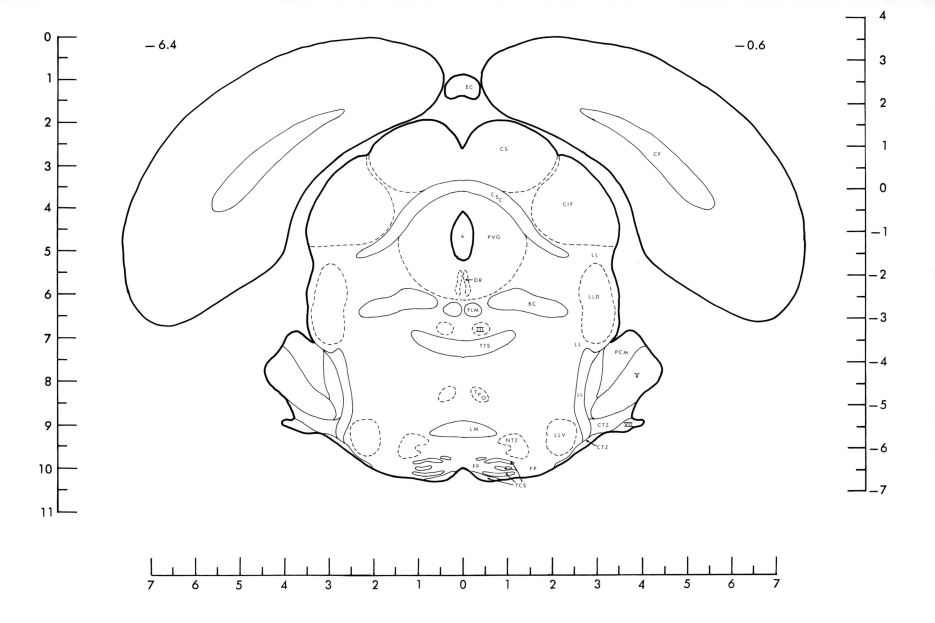

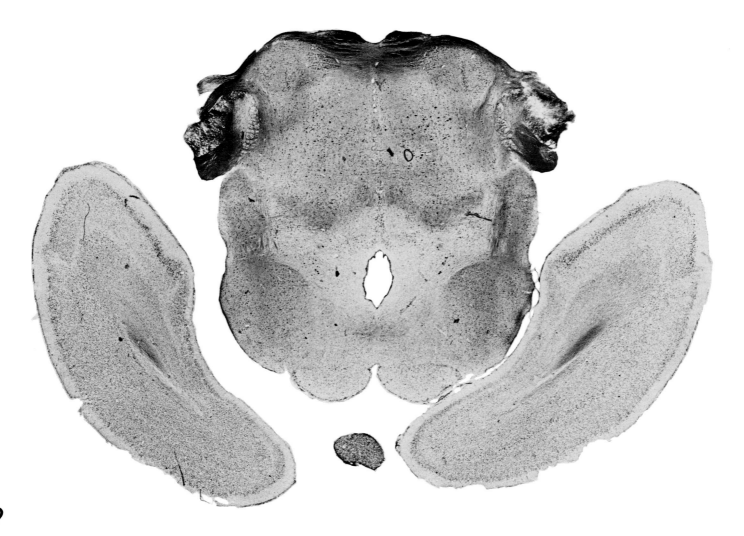

65

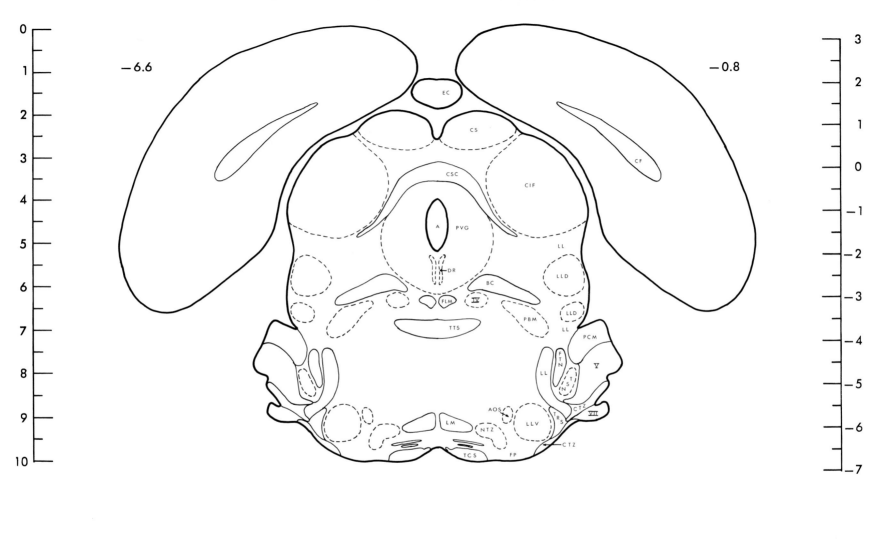

−6.6   −0.8

EC

CS

CF

CSC
CIF

A   PVG

LL

DR

LLD

BC

FLM   IV

LLD

PBM   LL

TTS   PCM

F
T
N   V

LL   T
S
N

AOS   CTZ

TRS   VII

LM   LLV

NTZ   CTZ

TCS   FP

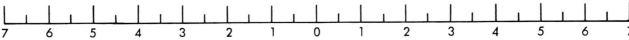

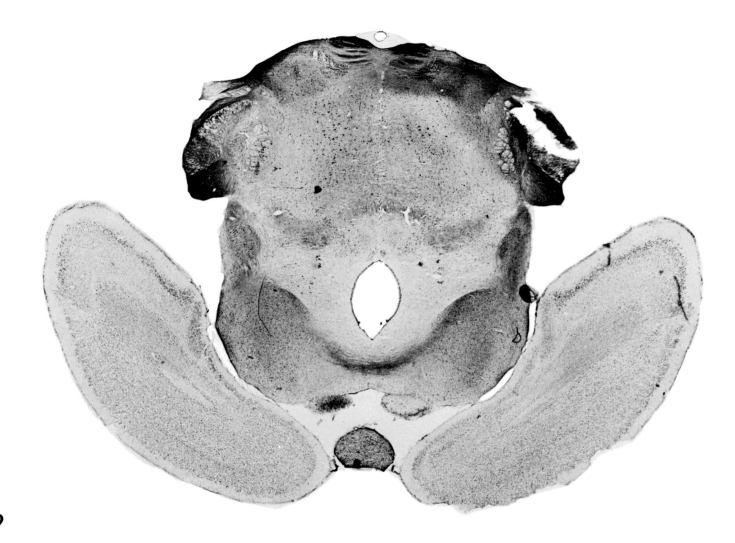

99

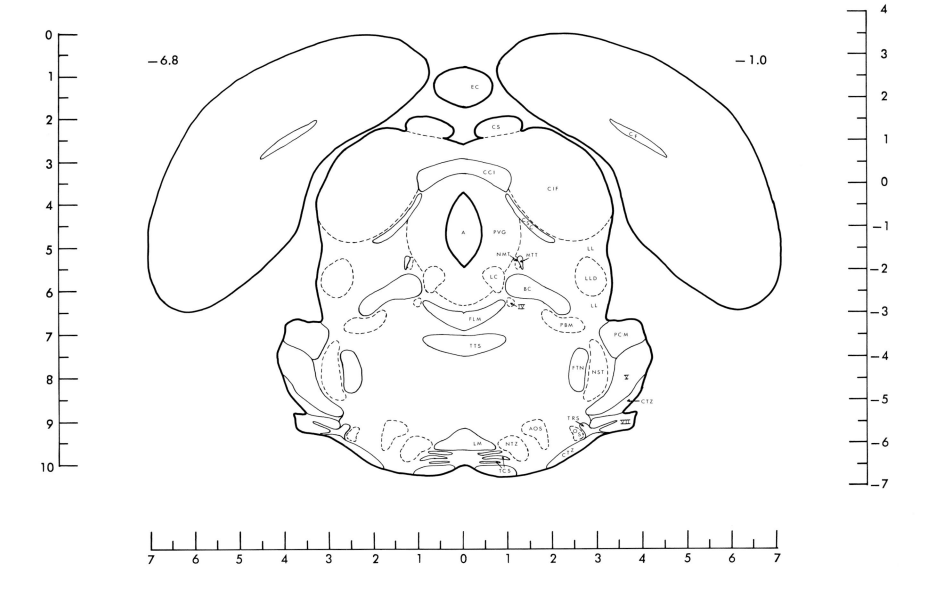

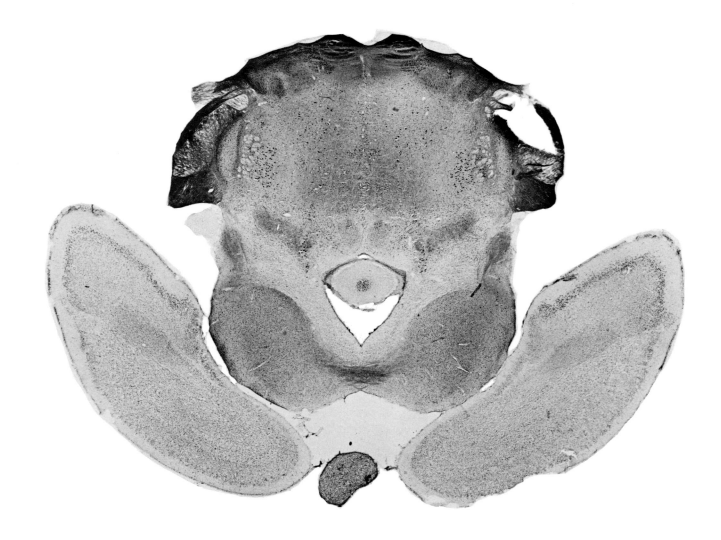

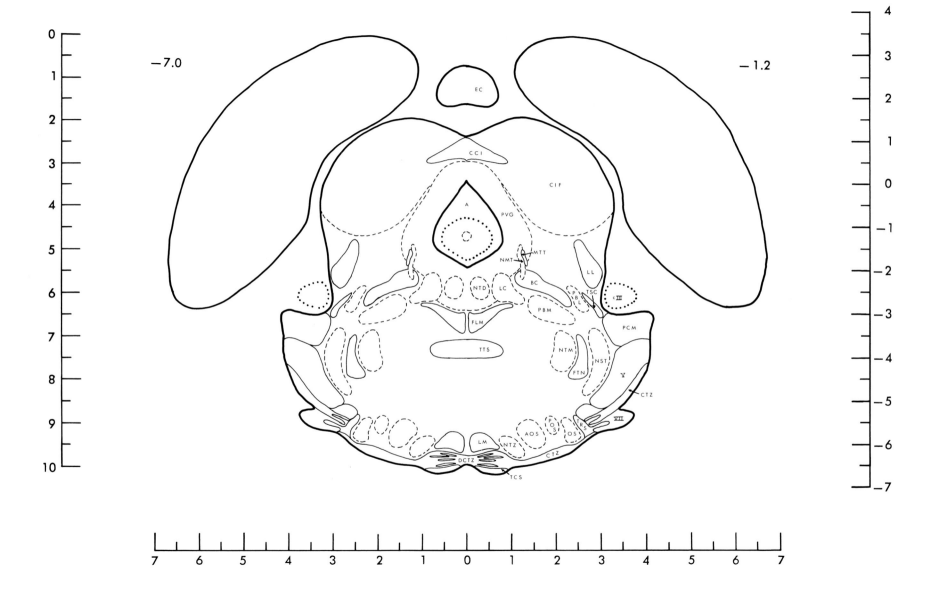

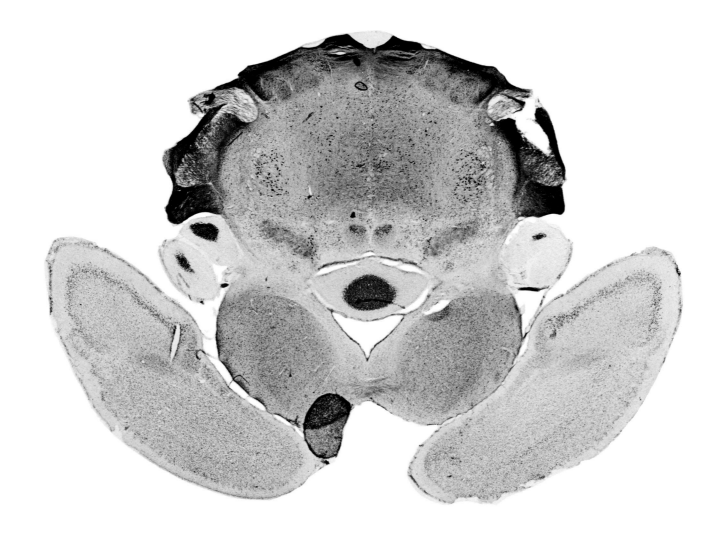

89

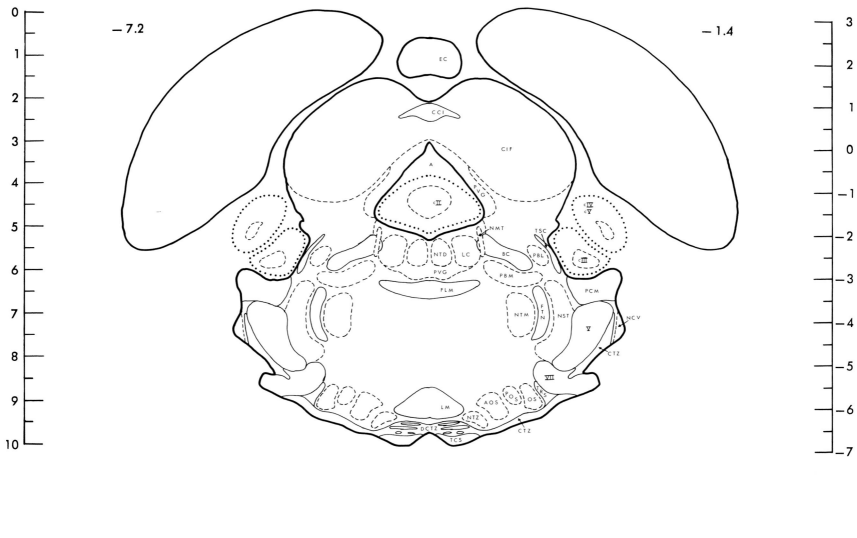
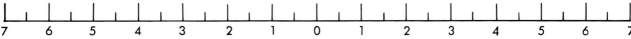

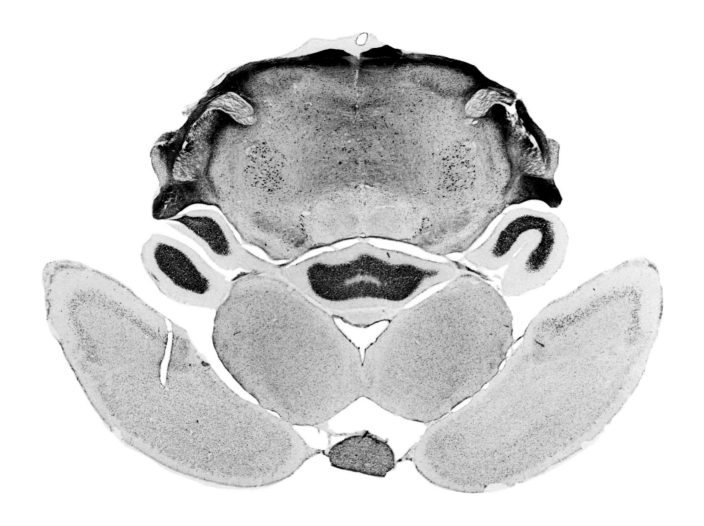

69

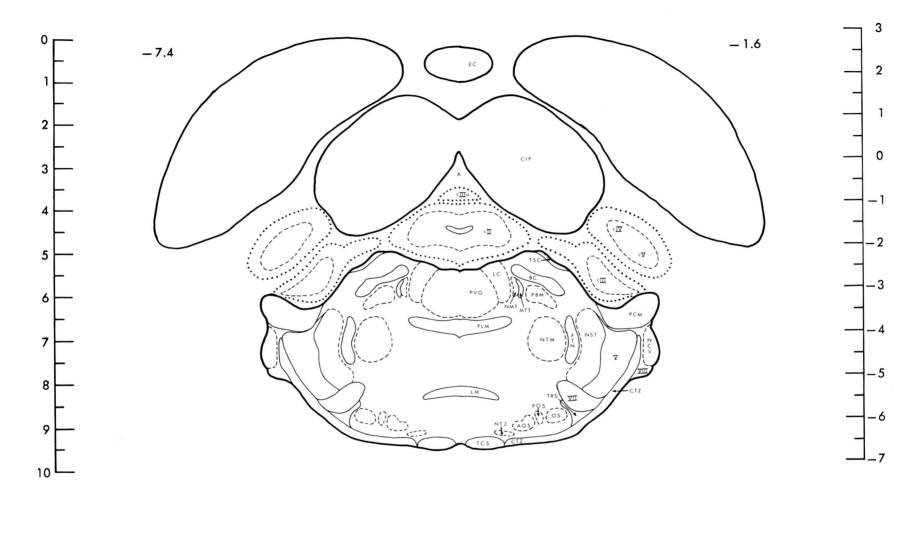

-7.4

-1.6

0
1
2
3
4
5
6
7
8
9
10

3
2
1
0
-1
-2
-3
-4
-5
-6
-7

EC

CIF

A

cIIIα

cII

cIV

cV

TSC

LC

BC

cIII

PVG

PBM

NMT MTT

FLM

PCM

NTM

F
T
N

NST

V

N
C
V

LM

VIII

TRS

VII

CTZ

POS

OS

NTZ

AOS

TCS

CTZ

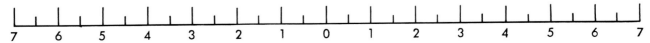

7   6   5   4   3   2   1   0   1   2   3   4   5   6   7

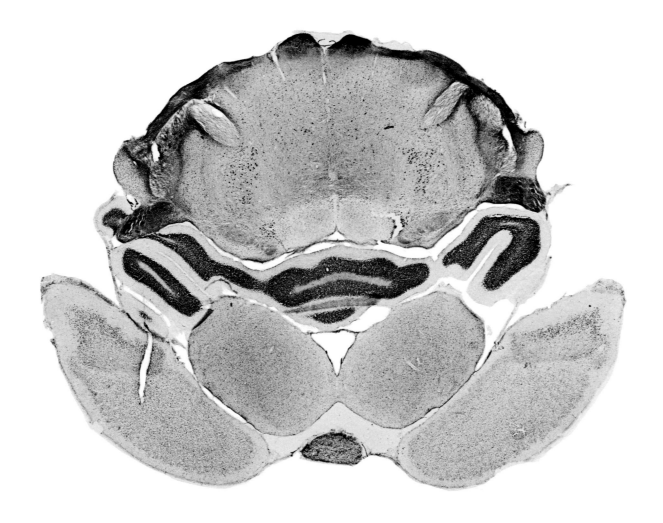

70

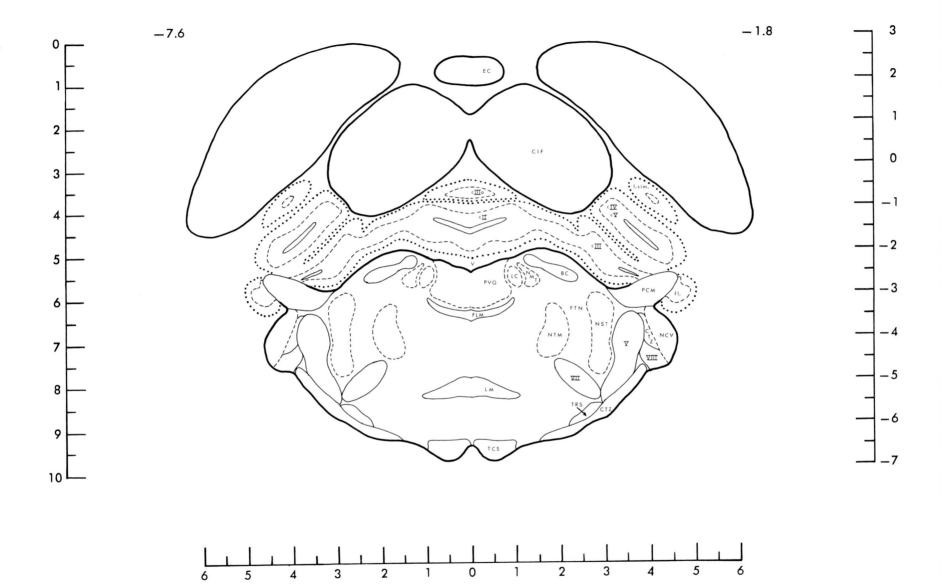

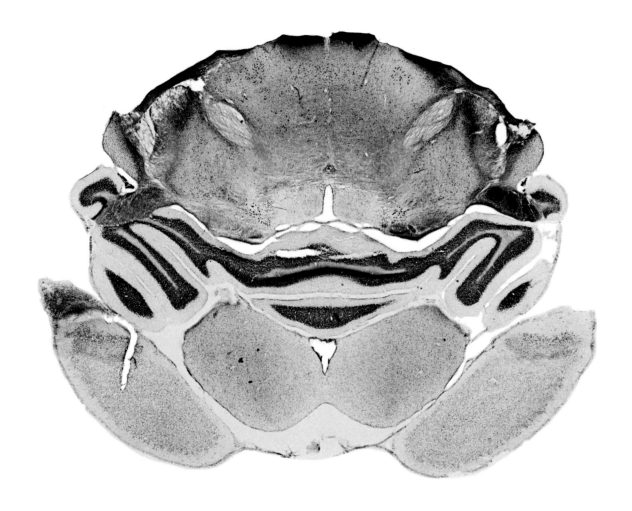

71

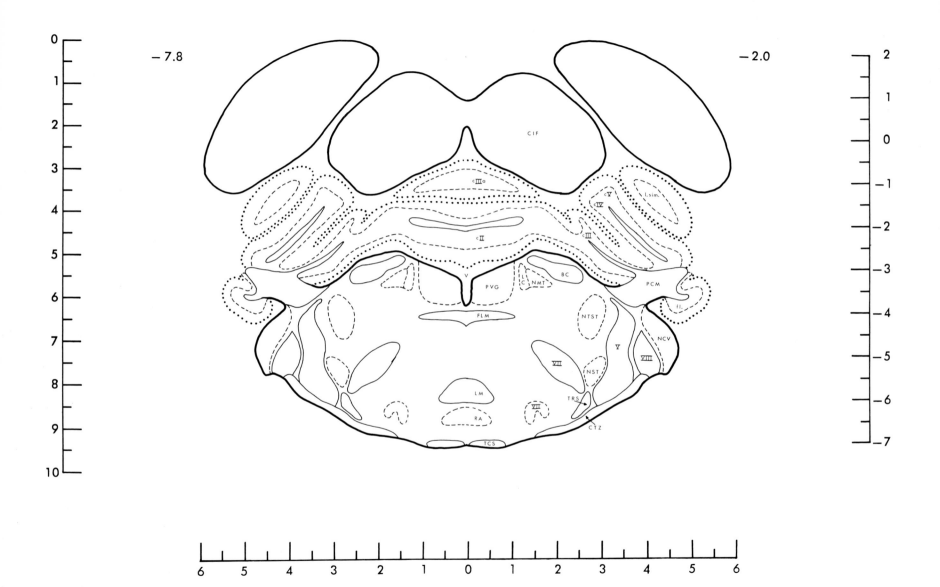

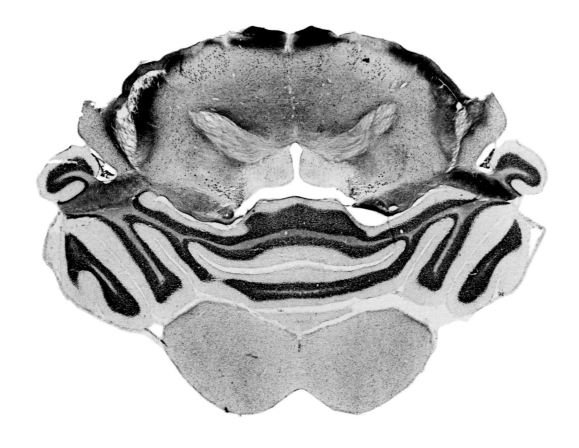

– 8.0

– 2.2

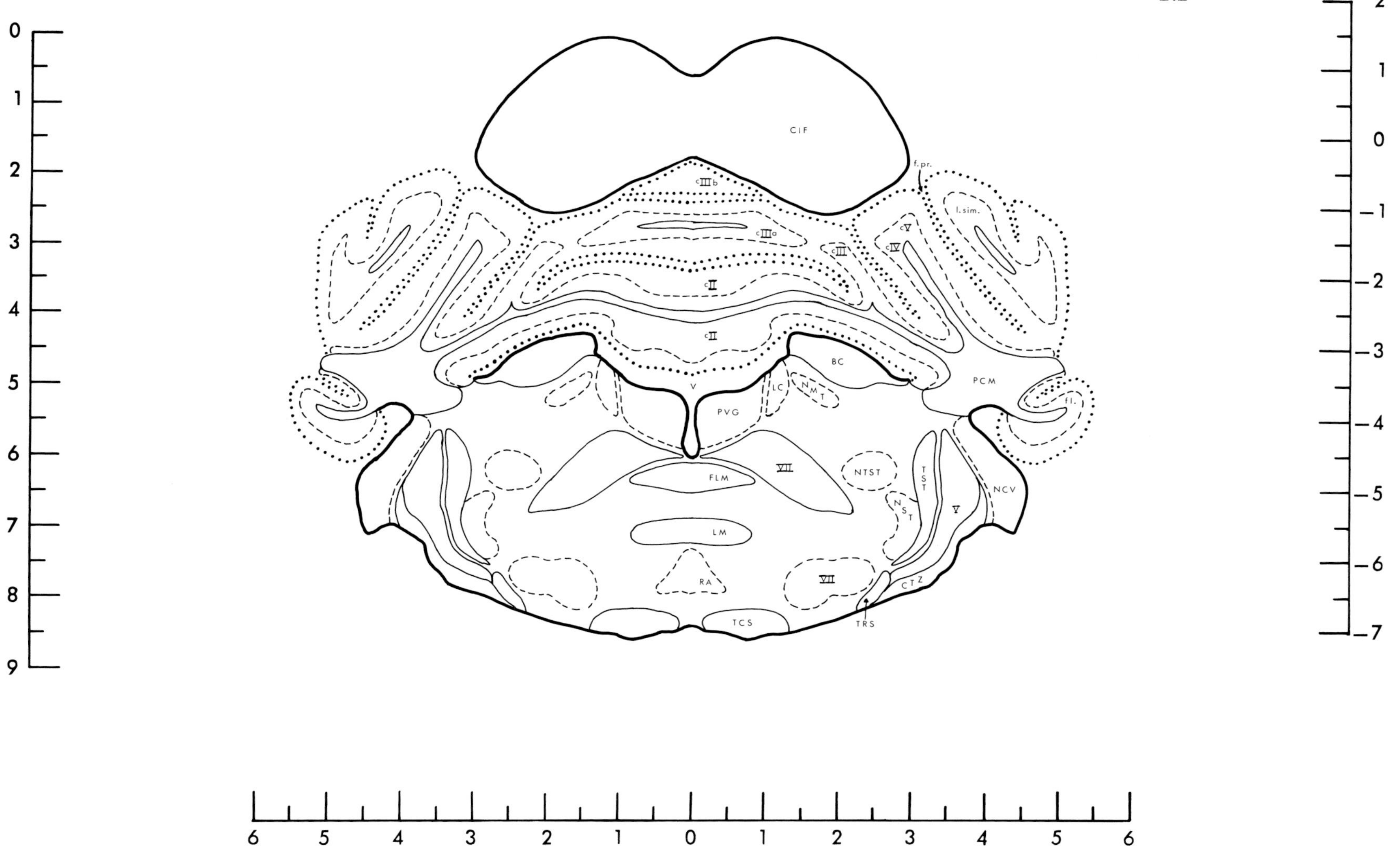
CIF
cIIIb
cIIIa
cIII
cII
f.pr.
l.sim.
cV
cIV
BC
V
LC
NMT
PVG
PCM
fl.
VII
NTST
TST
FLM
NST
V
NCV
LM
RA
VII
CTZ
TRS
TCS

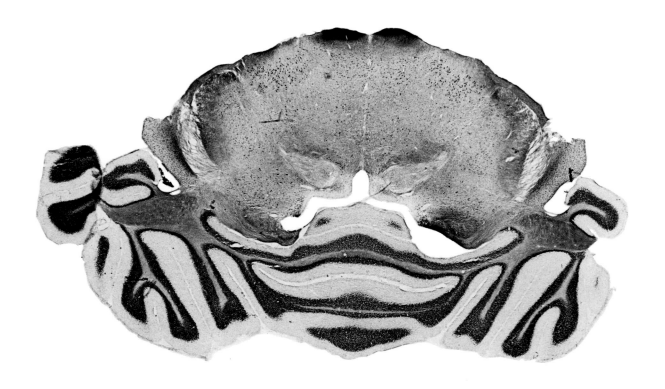

73

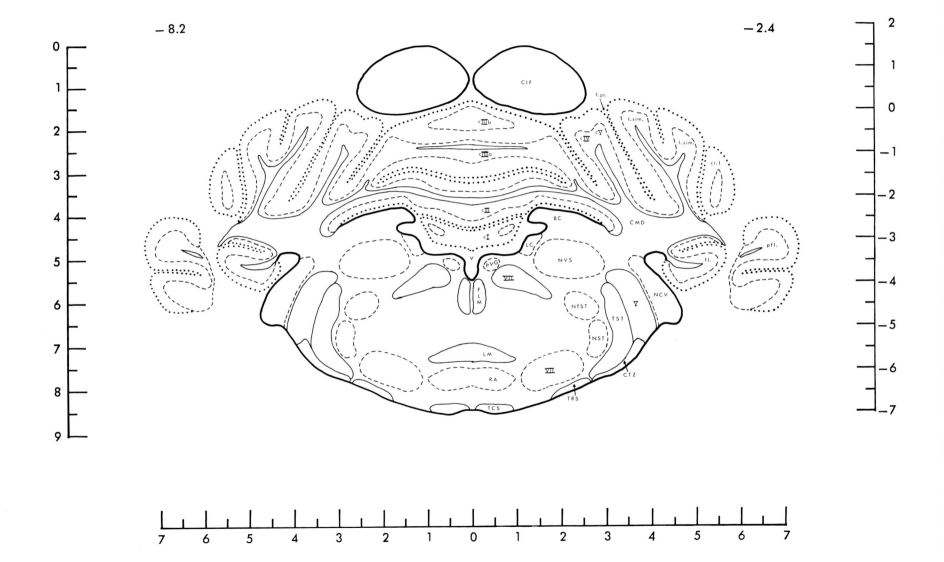

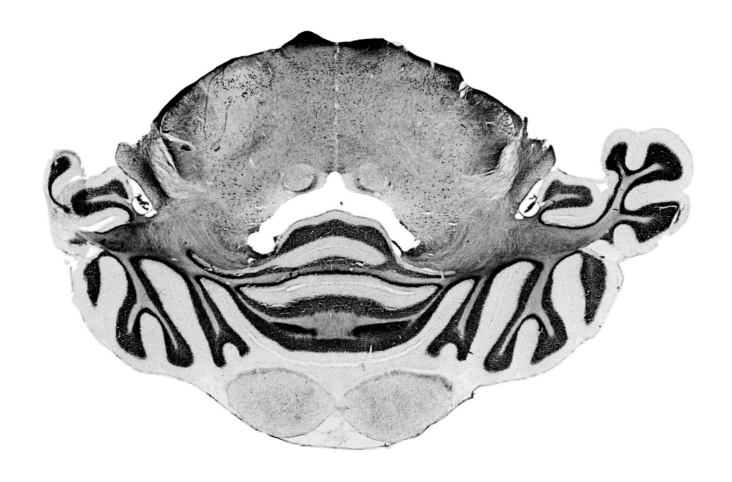

74

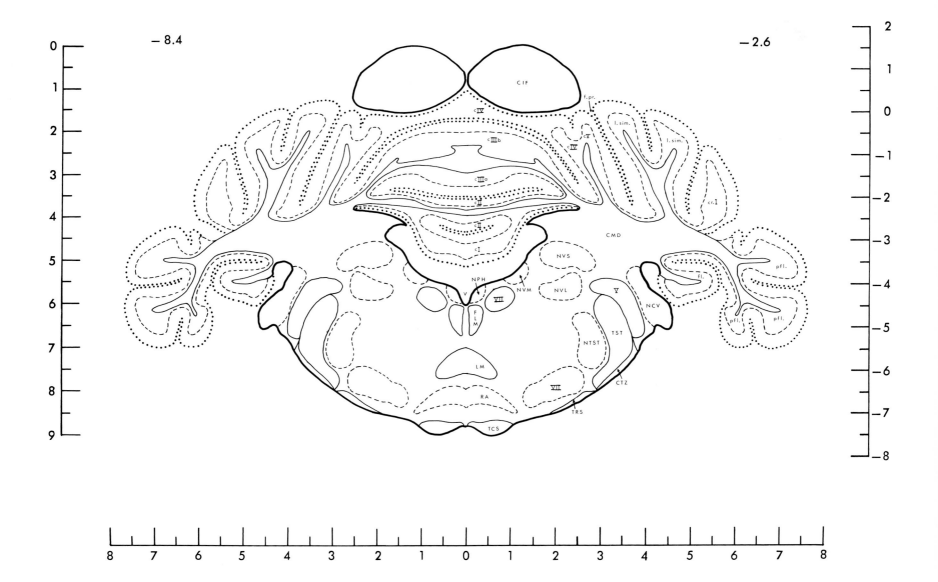

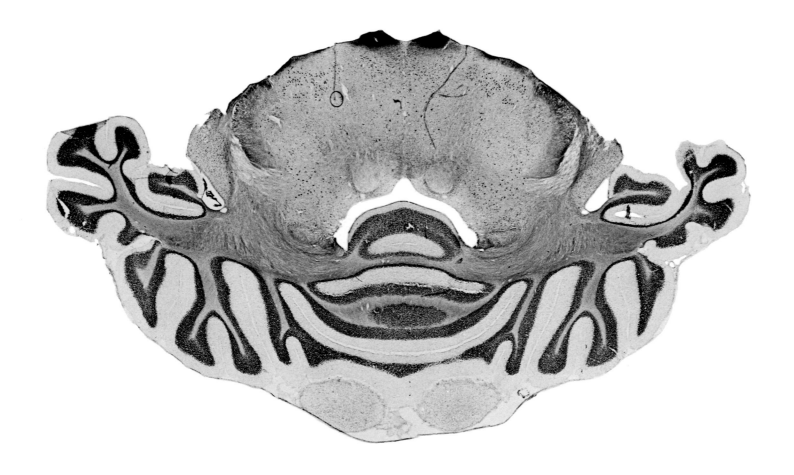

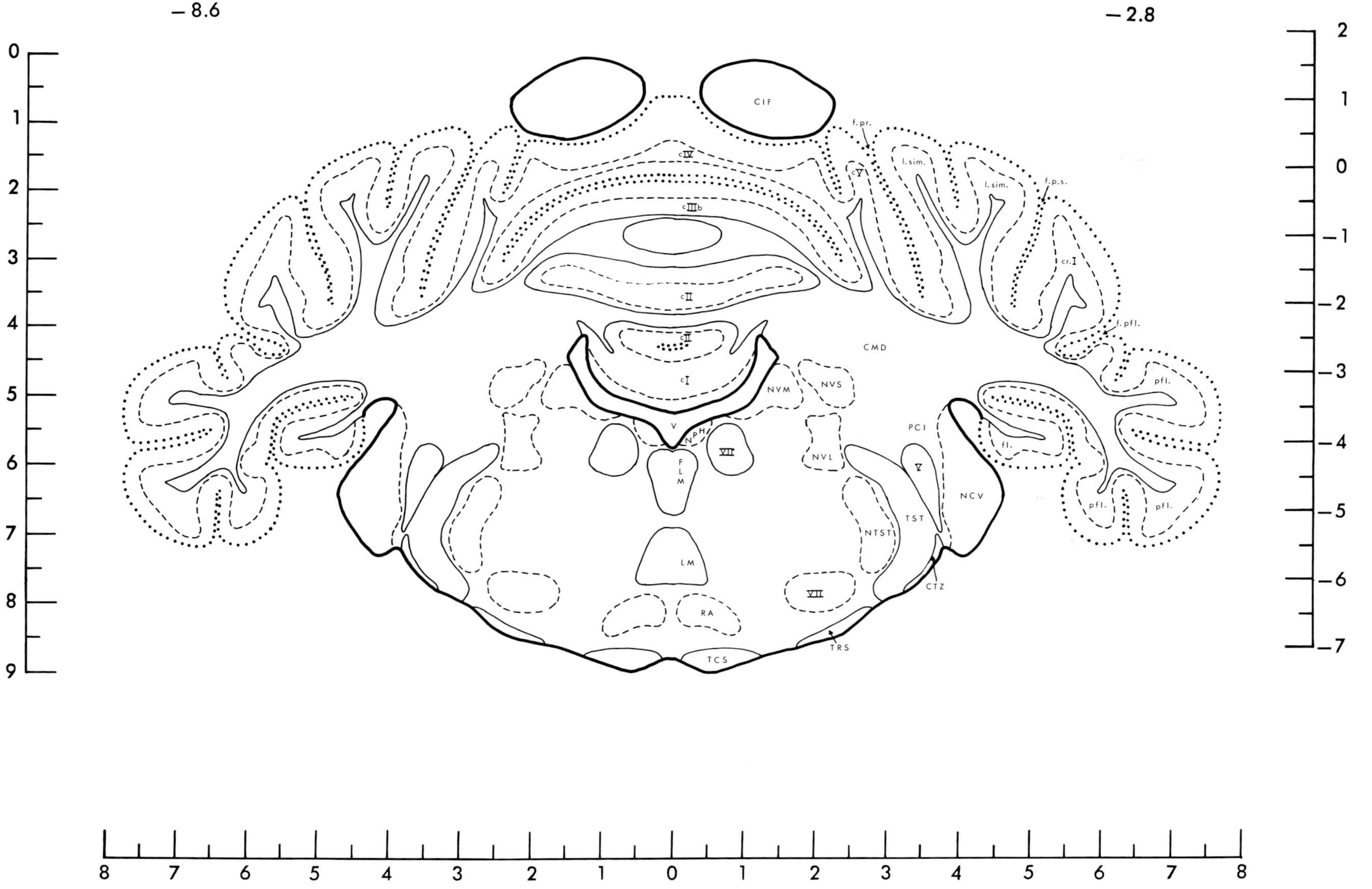

- 8.6
- 2.8
CIF
f. pr.
cIV
cV
l. sim.
l. sim.
f. p. s.
cIIIb
cr. I
cII
cII
CMD
f. pfl.
cI
NVM
NVS
pfl.
V
N
PH
VII
PCI
fl.
F
L
M
NVL
V
NCV
pfl.
pfl.
TST
NTST
LM
CTZ
VII
RA
TRS
TCS

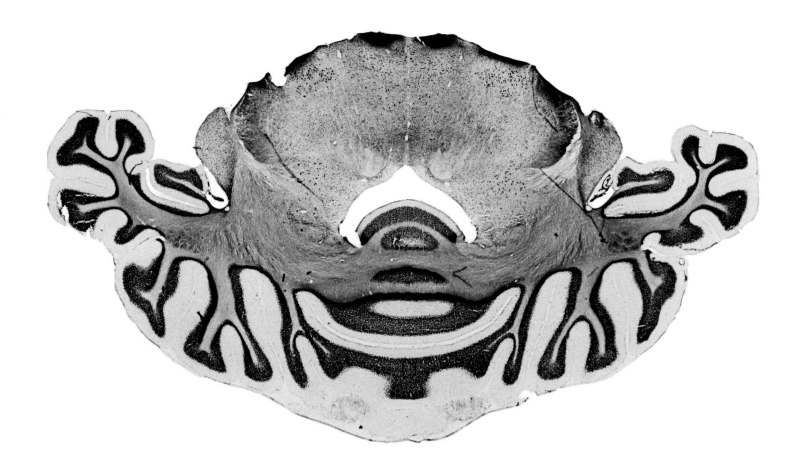

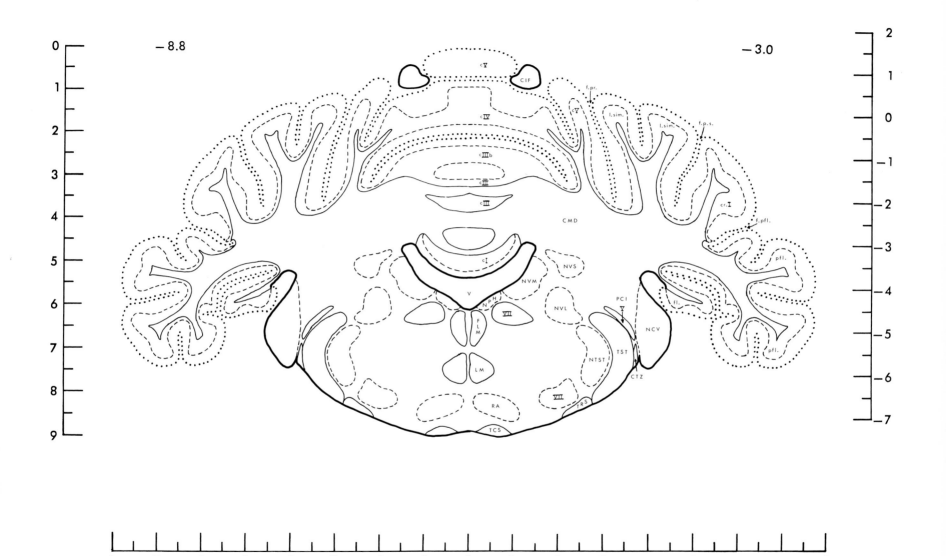

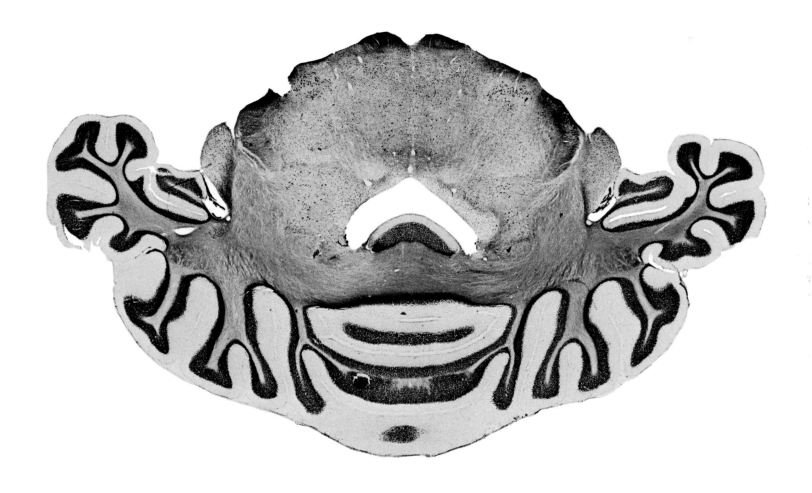

77

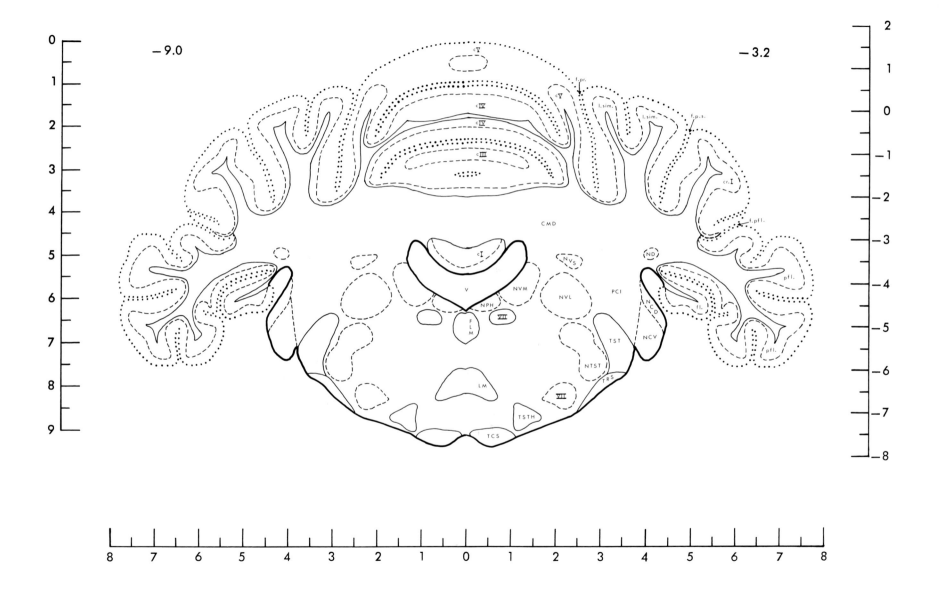

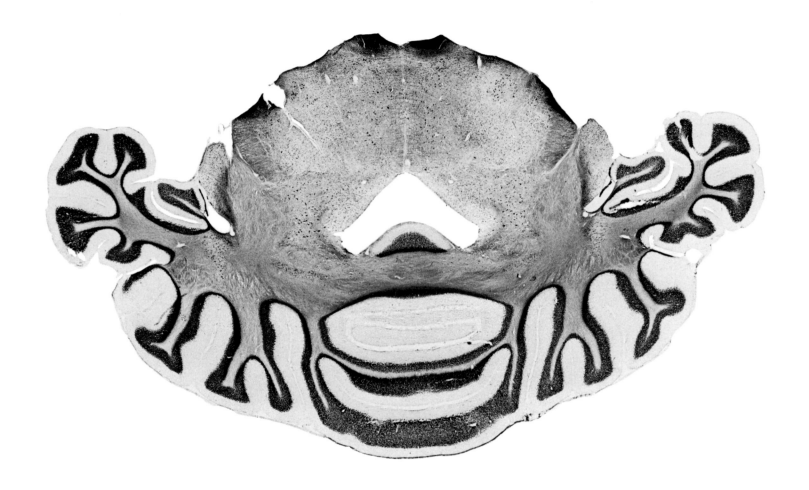

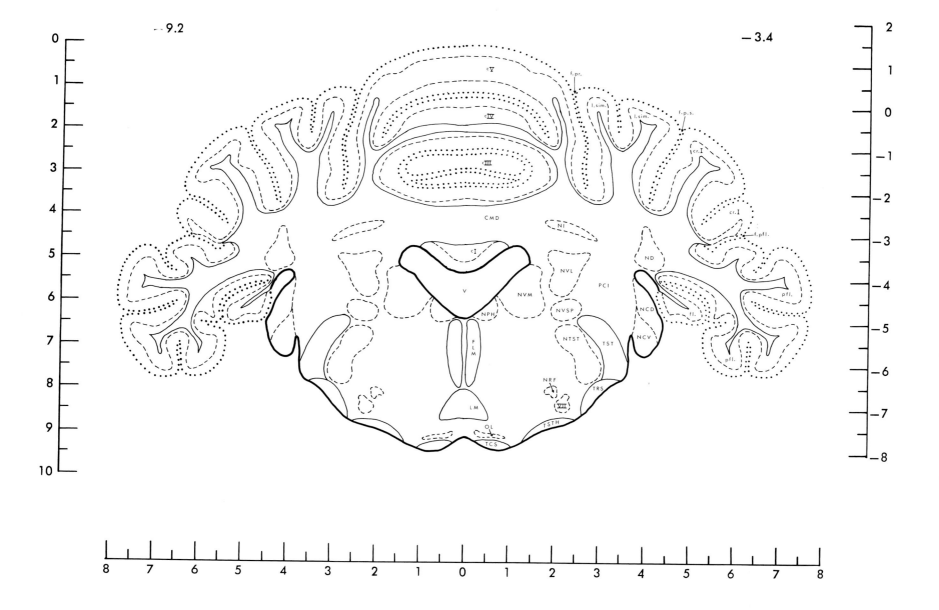

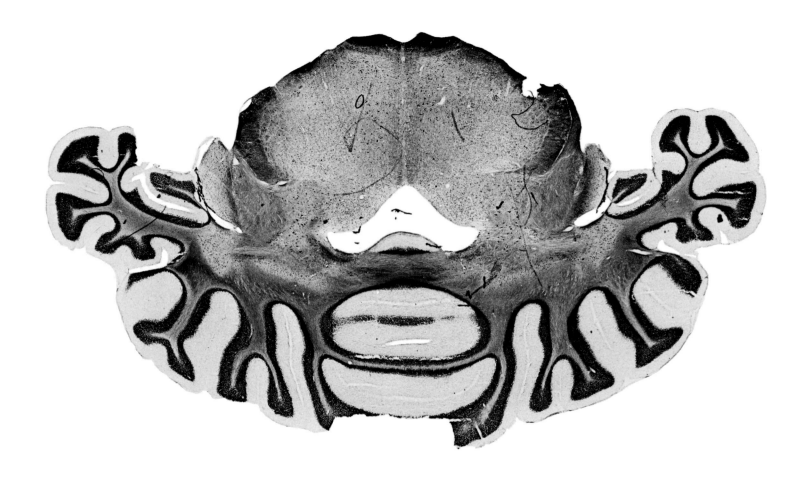

79

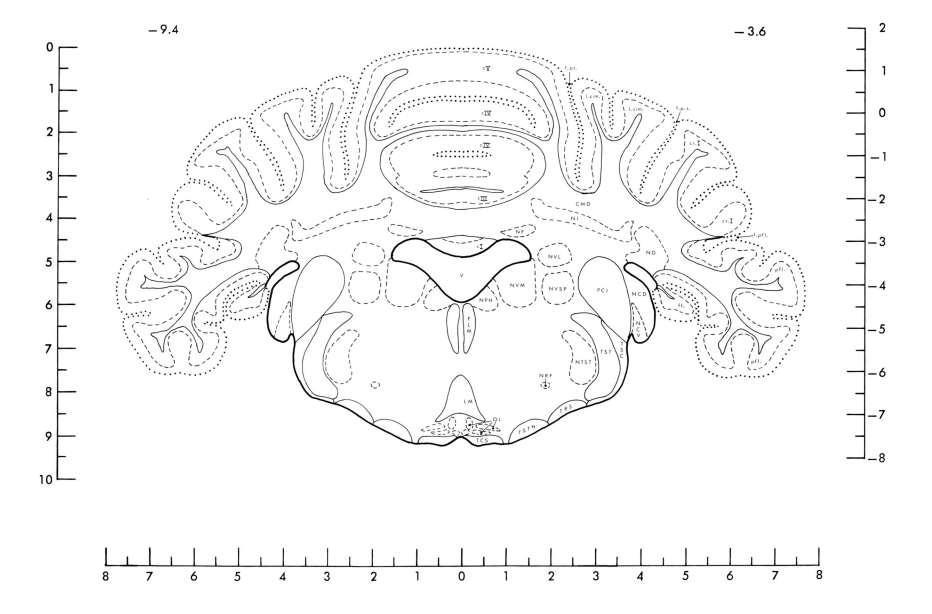

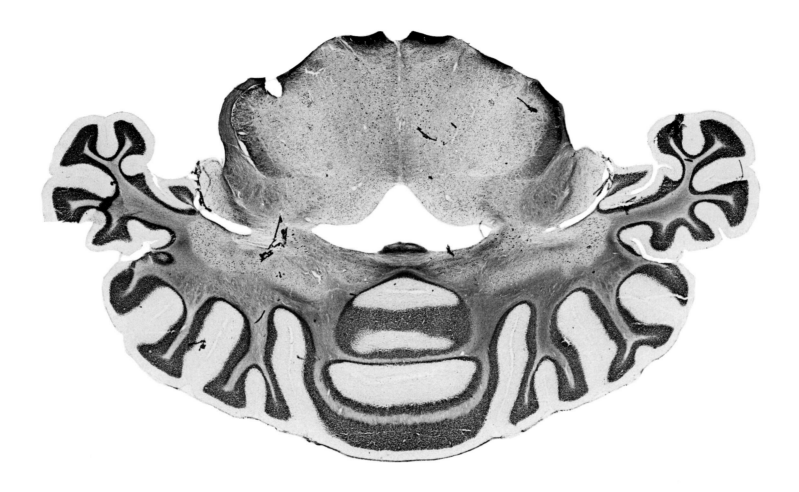

08

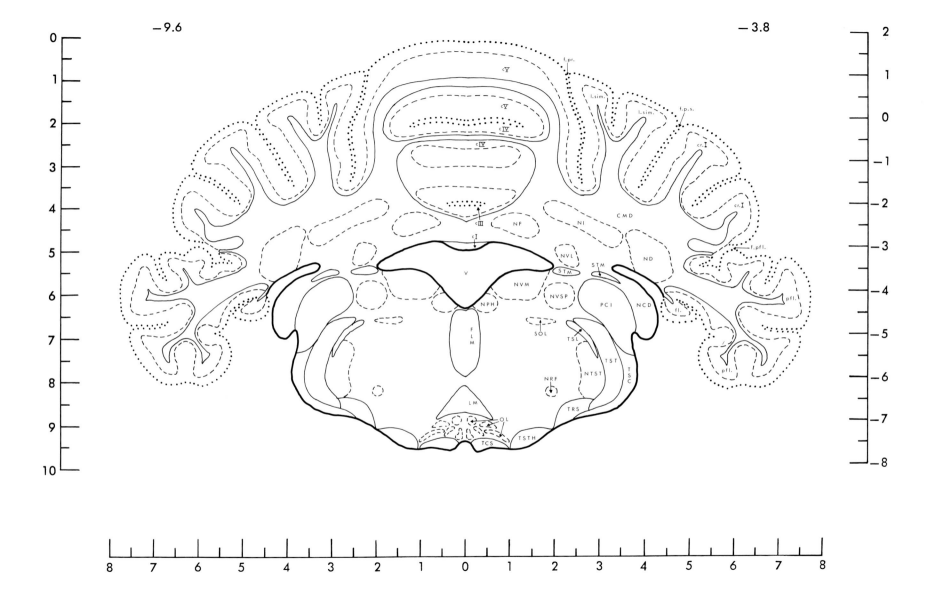

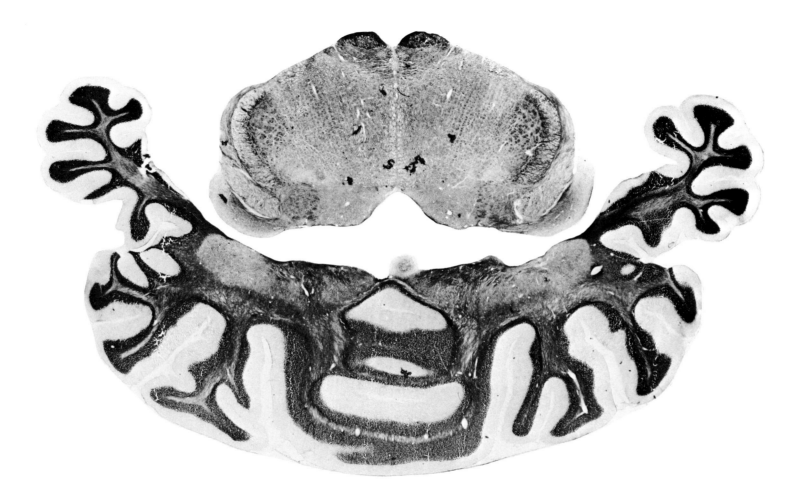

81

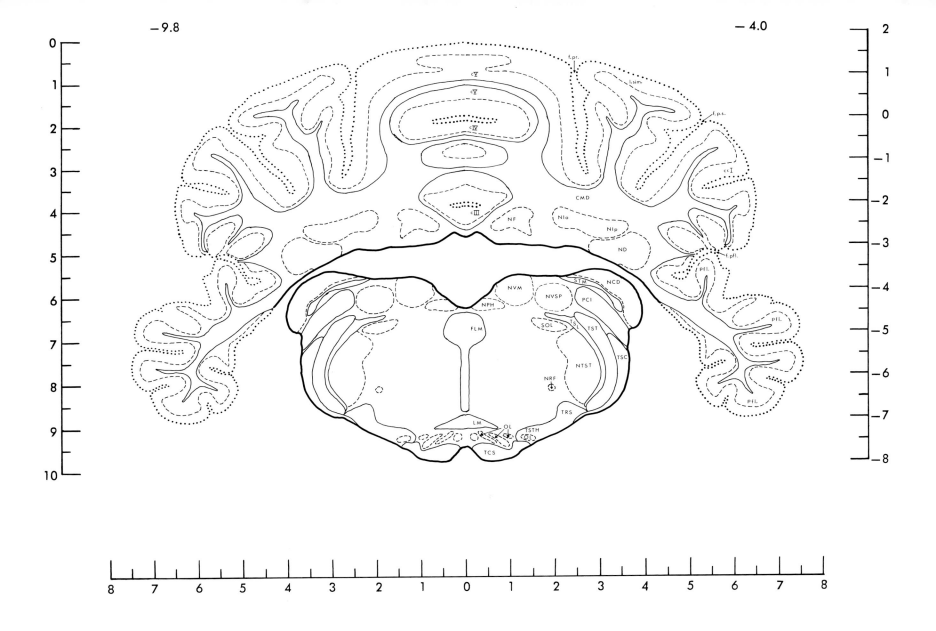

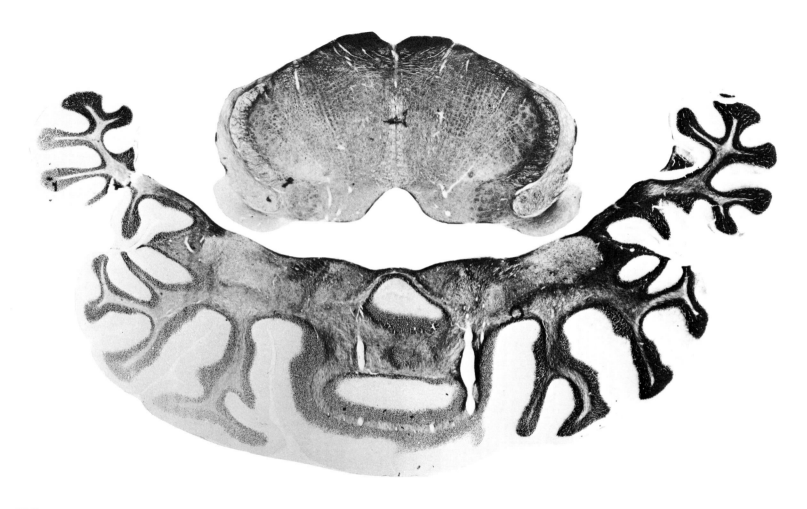

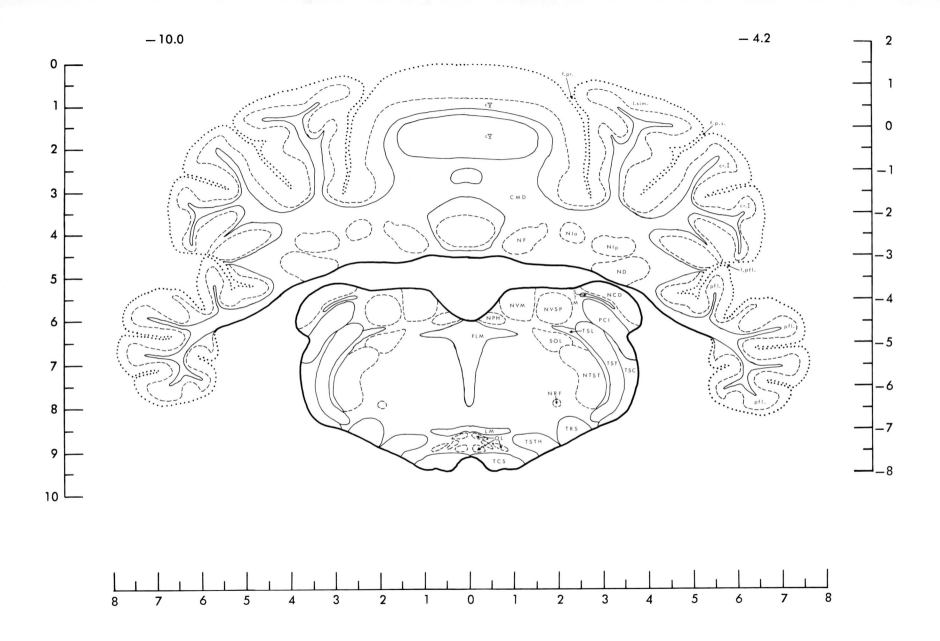

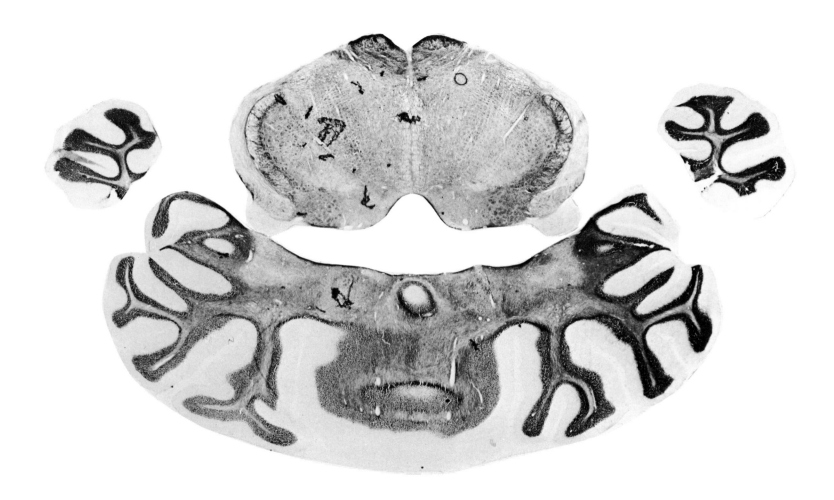

83

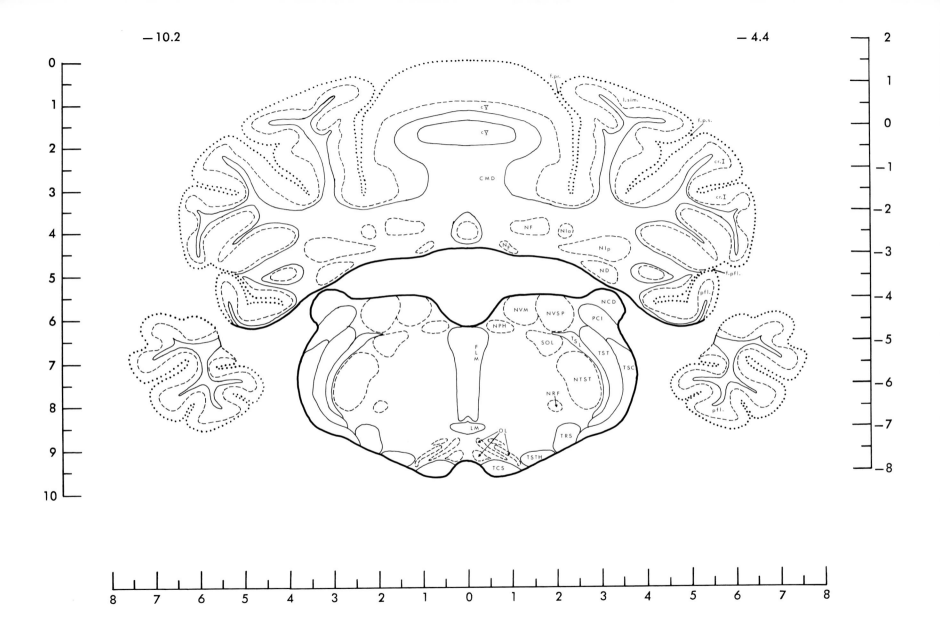

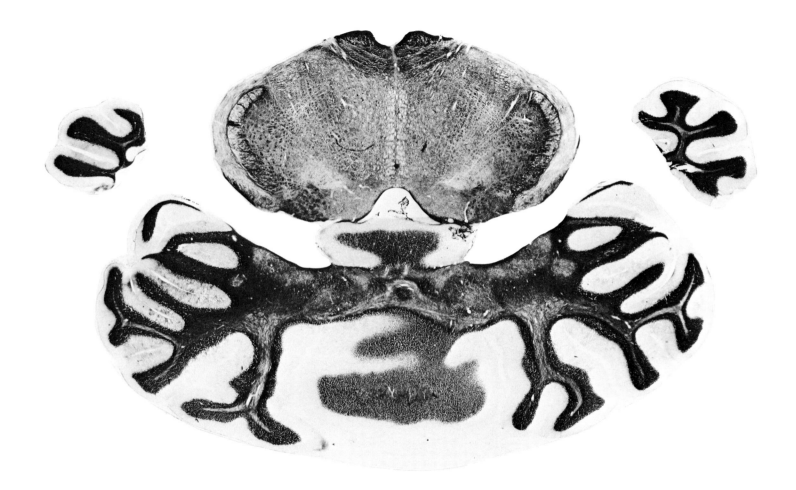

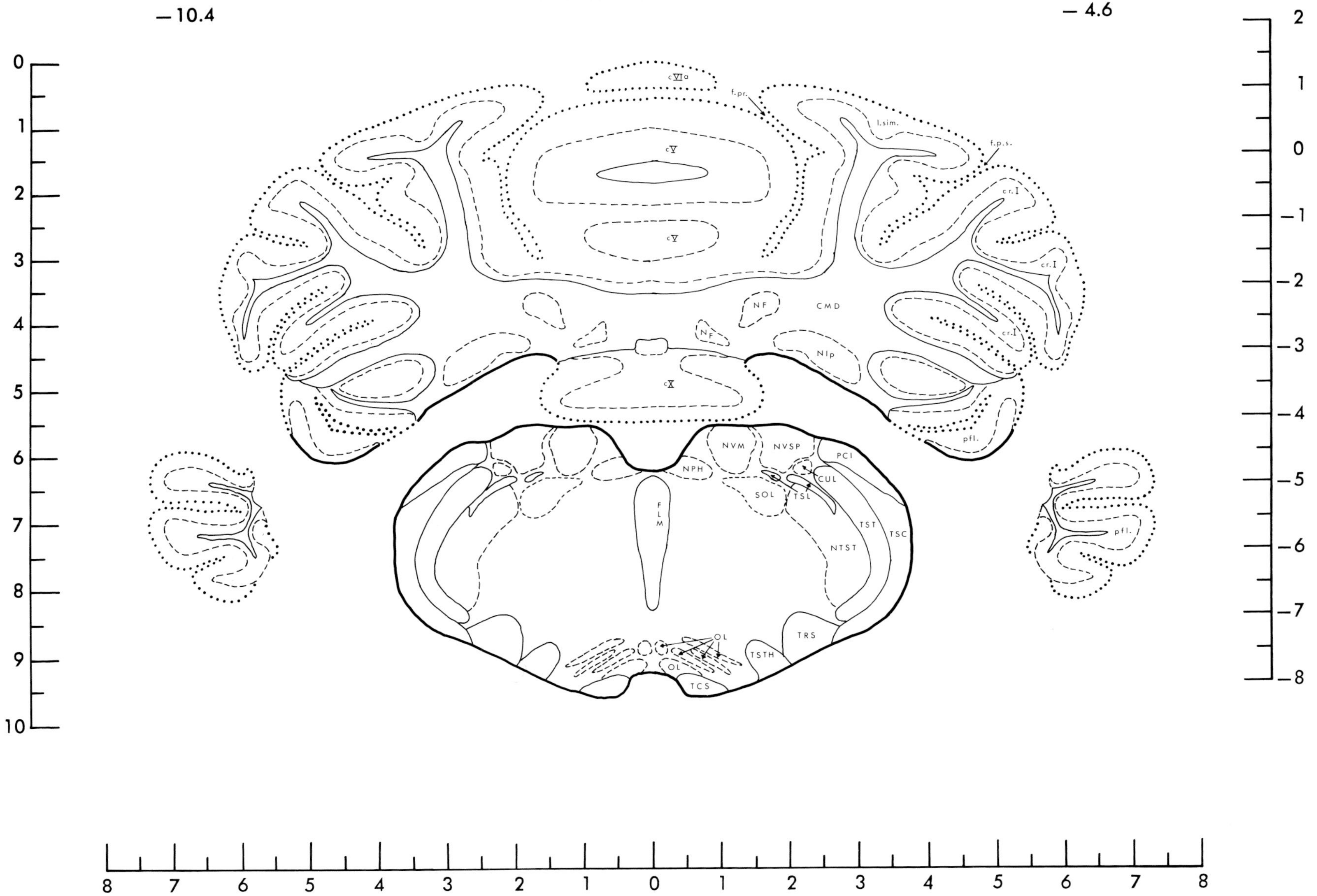

— 10.4
— 4.6
cVIa
f.pr.
l.sim.
f.p.s.
cV
cV
cr.I
cr.I
cr.I
NF
CMD
NF
NIp
cX
pfl.
NVM
NVSP
PCI
NPH
CUL
SOL
TSL
F L M
TST
TSC
NTST
pfl.
TRS
OL
TSTH
OL
TCS
0
1
2
3
4
5
6
7
8
9
10
2
1
0
−1
−2
−3
−4
−5
−6
−7
−8
8
7
6
5
4
3
2
1
0
1
2
3
4
5
6
7
8

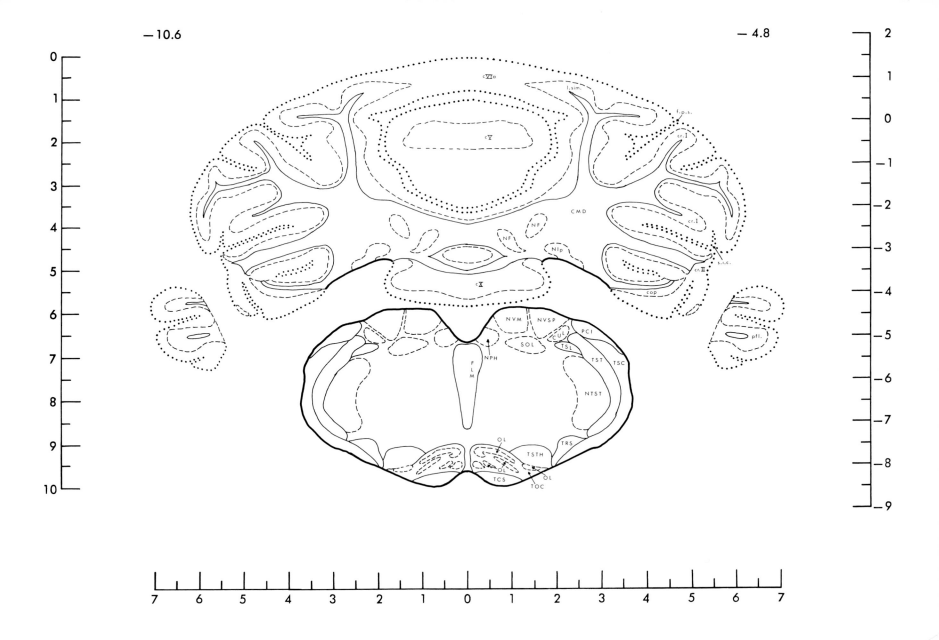

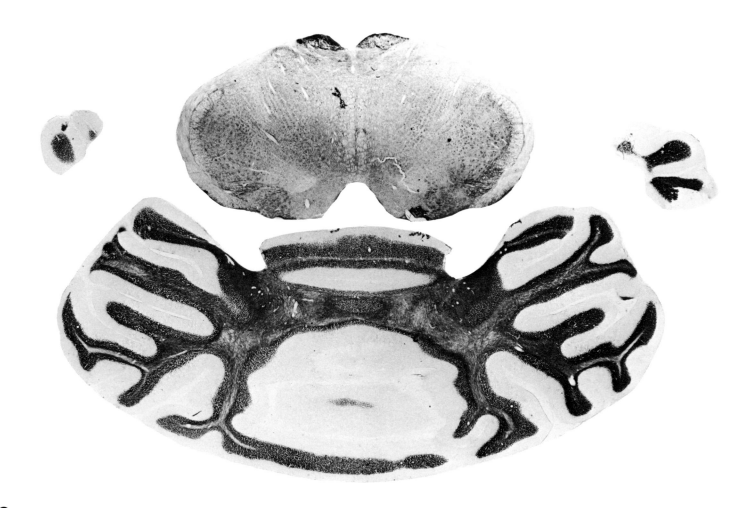

98

−10.8

−5.0

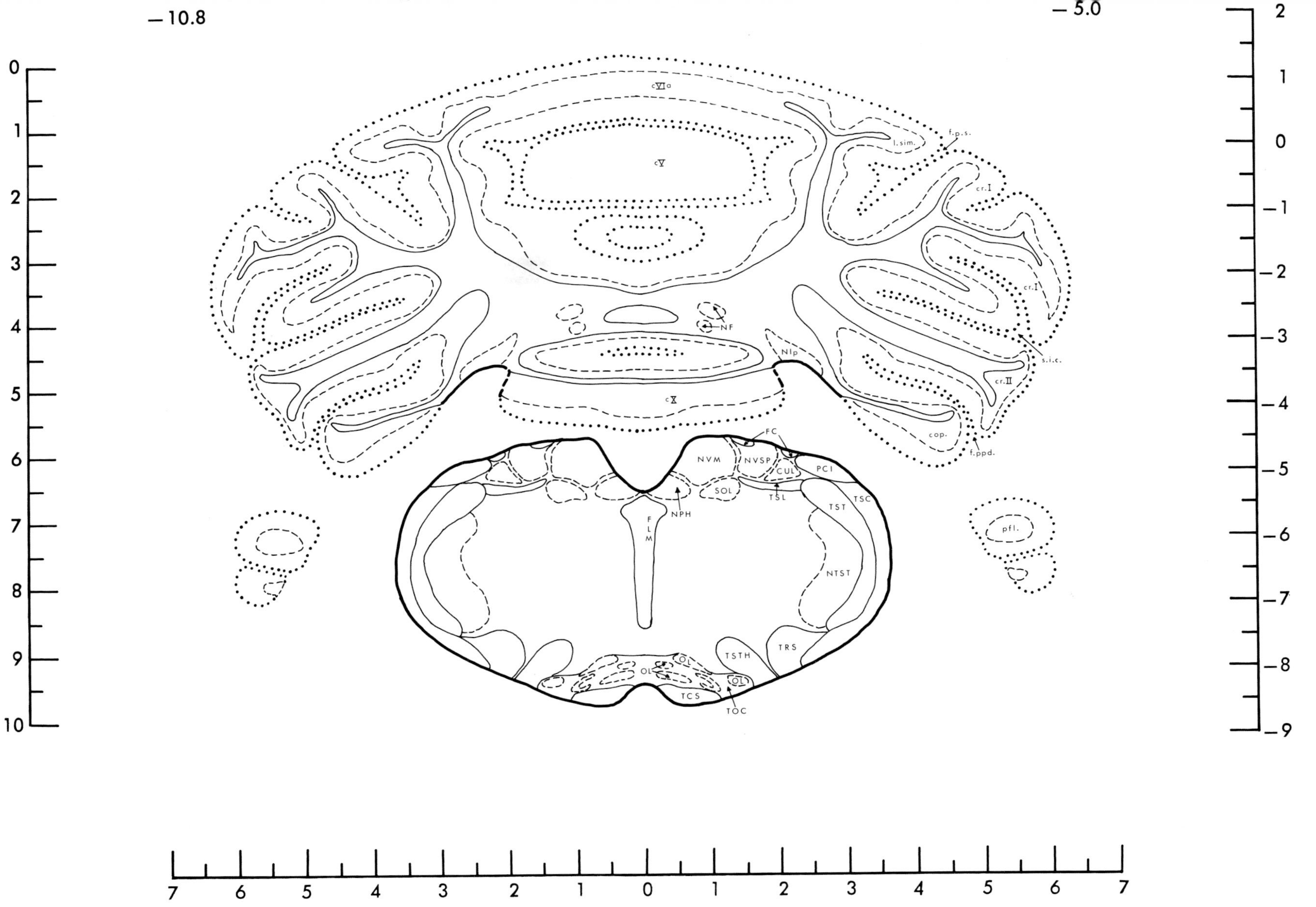

cVIa
cV
l.sim.
f.p.s.
cr.I
cr.II
s.i.c.
cr.II
cop.
f.ppd.
NF
NIp
cX
FC
NVM
NVSP
CUL
PCI
SOL
TSL
NPH
TST
TSC
F
L
M
pfl.
NTST
TRS
TSTH
OL
OL
OL
TCS
TOC

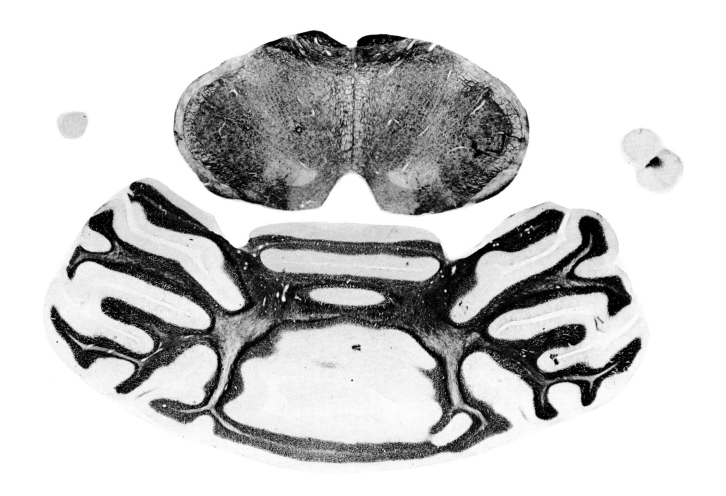

87

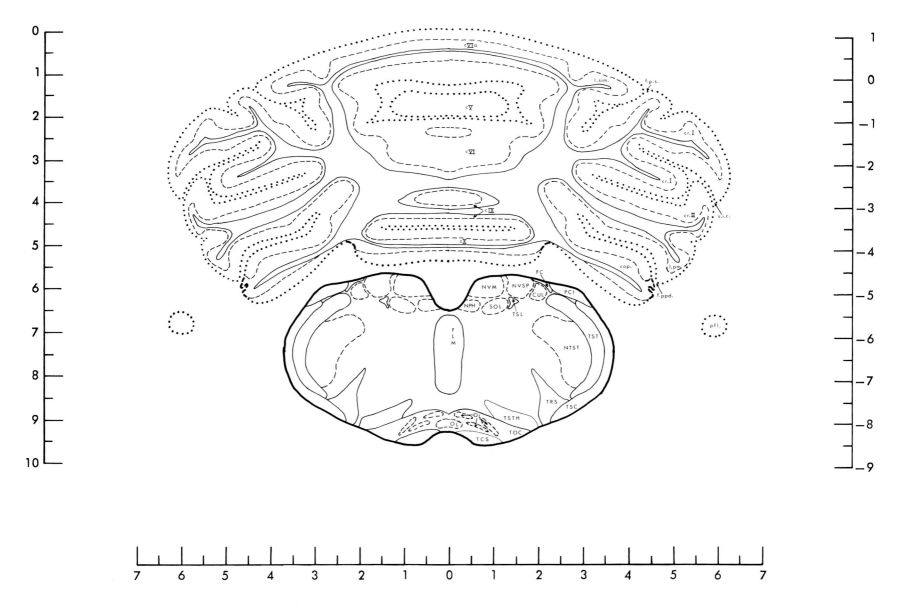

88

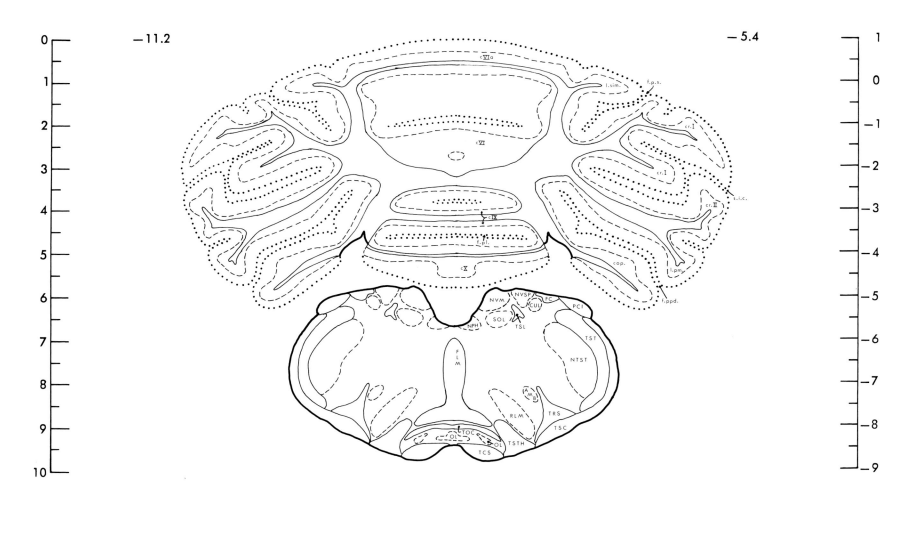

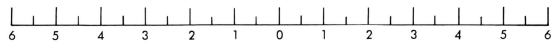

68

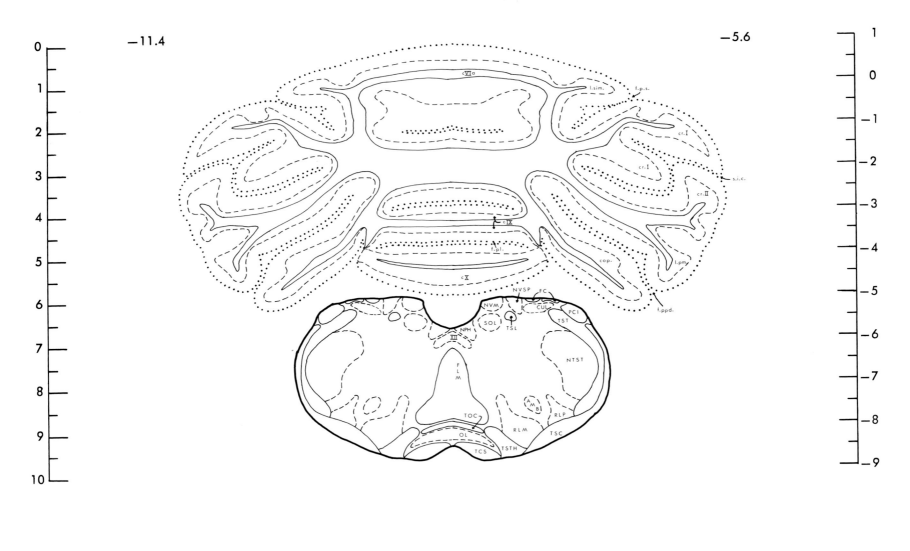

−11.4

−5.6

c VI a

l.sim.

f.p.s.

cr. I

cr. I

s.i.c.

cr. II

c IX

f.pl.

cop.

l.pm.

c X

f.ppd.

NVSP   FC

NVM   CUL

SOL   PCI

NPH   TSL   TST

XII

F
L
M

NTST

TOC

A
M
B

RLP

RLM

TSC

OL

TCS   TSTH

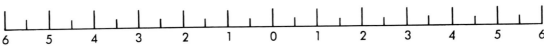

6   5   4   3   2   1   0   1   2   3   4   5   6

06

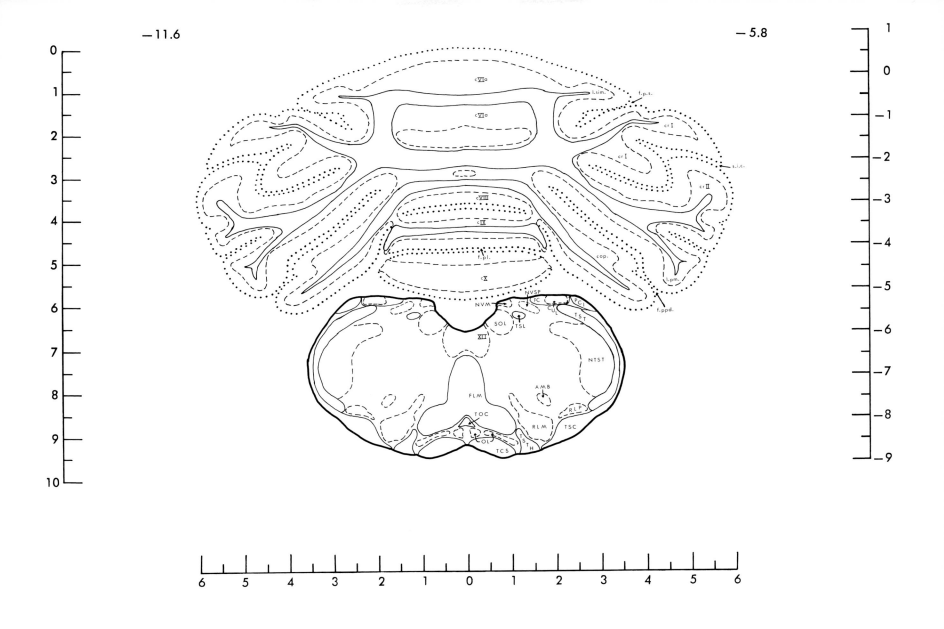

16

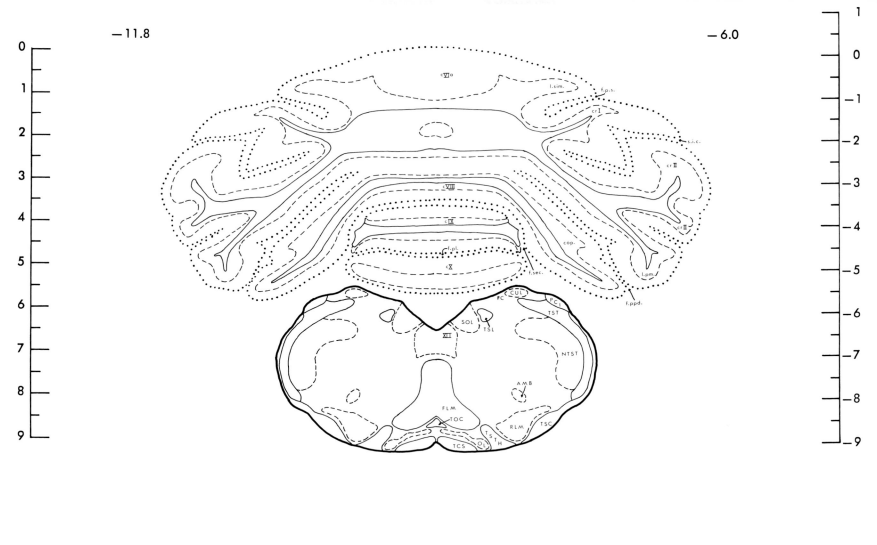

−11.8   −6.0

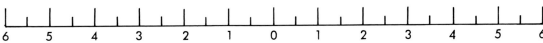

92

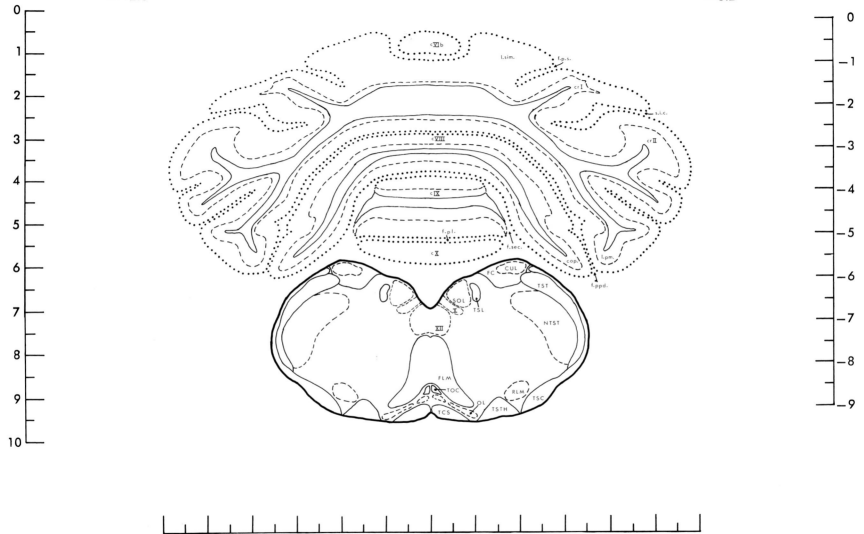

93

−12.2 −6.4 0

−1

−2

−3

−4

−5

−6

−7

−8

−9

0

1

2

3

4

5

6

7

8

9

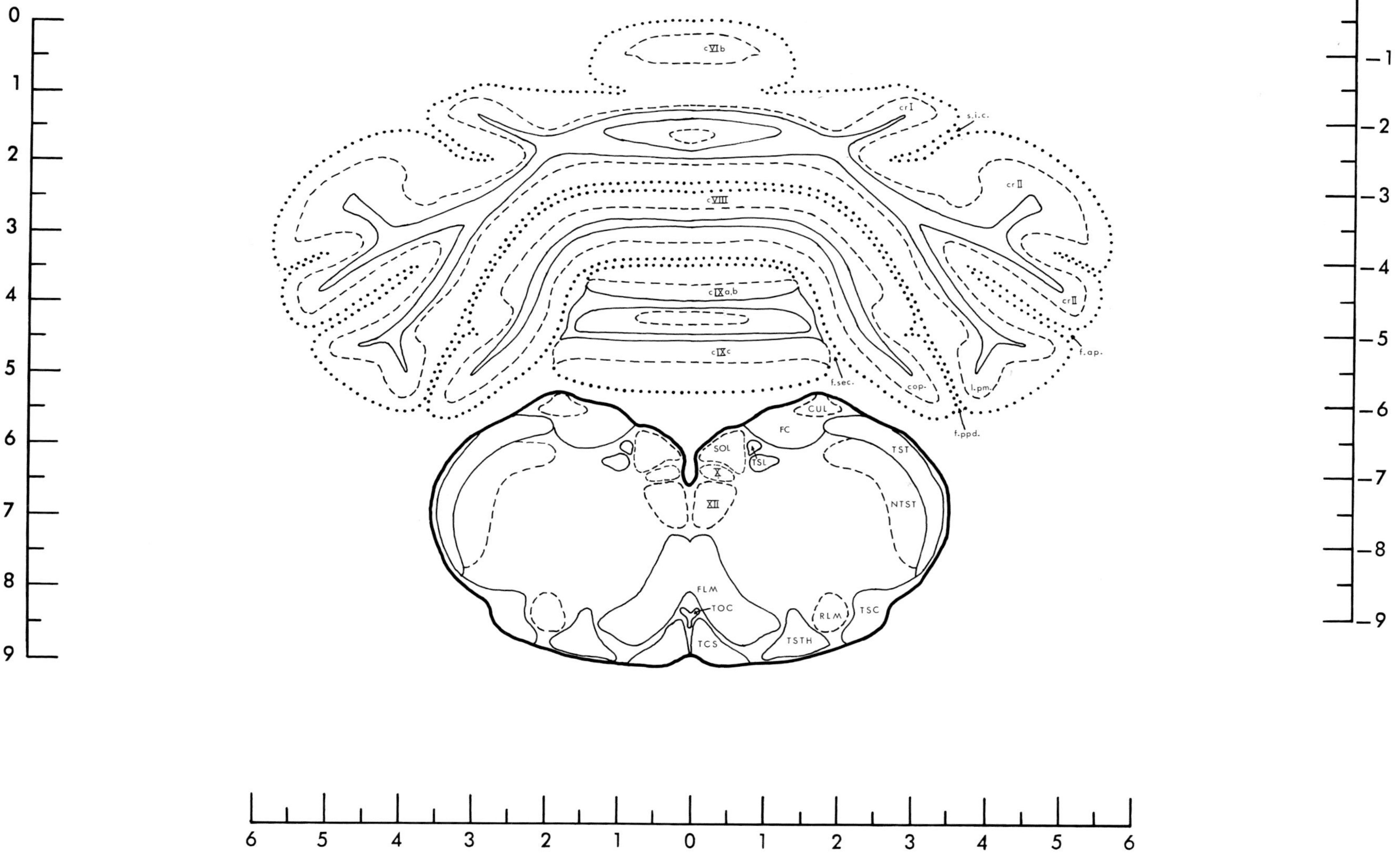

cⅥb

crⅠ

s.i.c.

crⅡ

cⅧ

crⅢ

cⅨa,b

f.ap.

cⅨc

f.sec. cop. l.pm.

f.ppd.

CUL

FC

SOL

TSL

TST

Ⅹ

XII

NTST

FLM

TOC

RLM TSC

TCS

TSTH

6 5 4 3 2 1 0 1 2 3 4 5 6

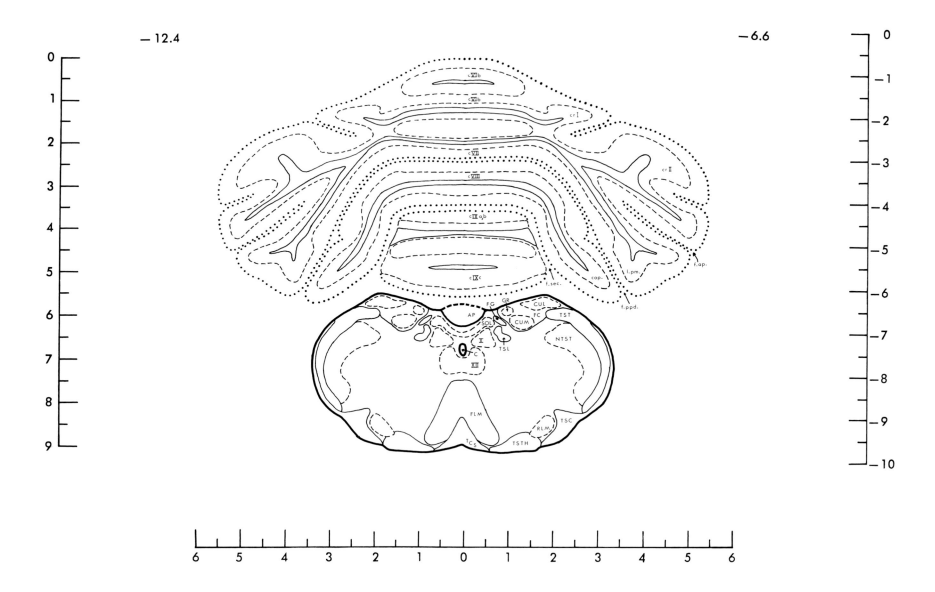

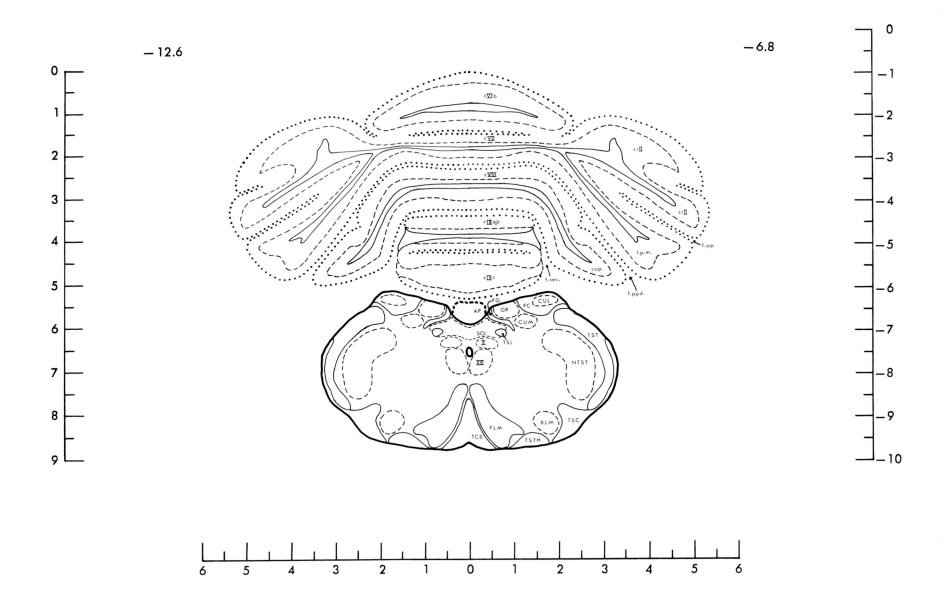

96

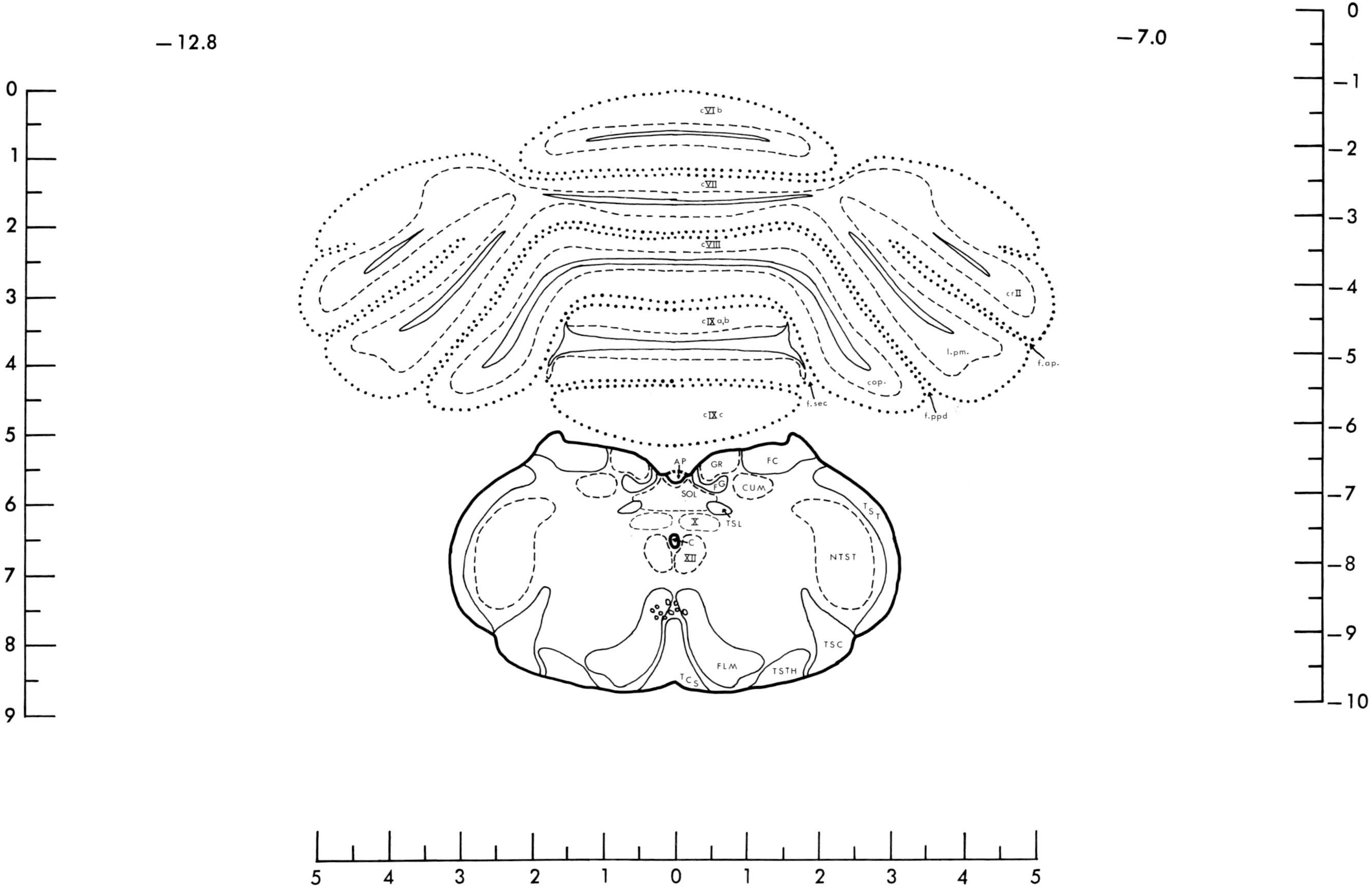

−12.8
−7.0
cVIb
cVII
cVIII
cIX a,b
cIX c
crII
l.pm.
cop.
f.ap.
f.sec
f.ppd
AP
GR
FC
FG
SOL
CUM
TST
X
TSL
C
XII
NTST
TSC
FLM
TSTH
TCS
0
1
2
3
4
5
6
7
8
9
0
−1
−2
−3
−4
−5
−6
−7
−8
−9
−10
5 4 3 2 1 0 1 2 3 4 5

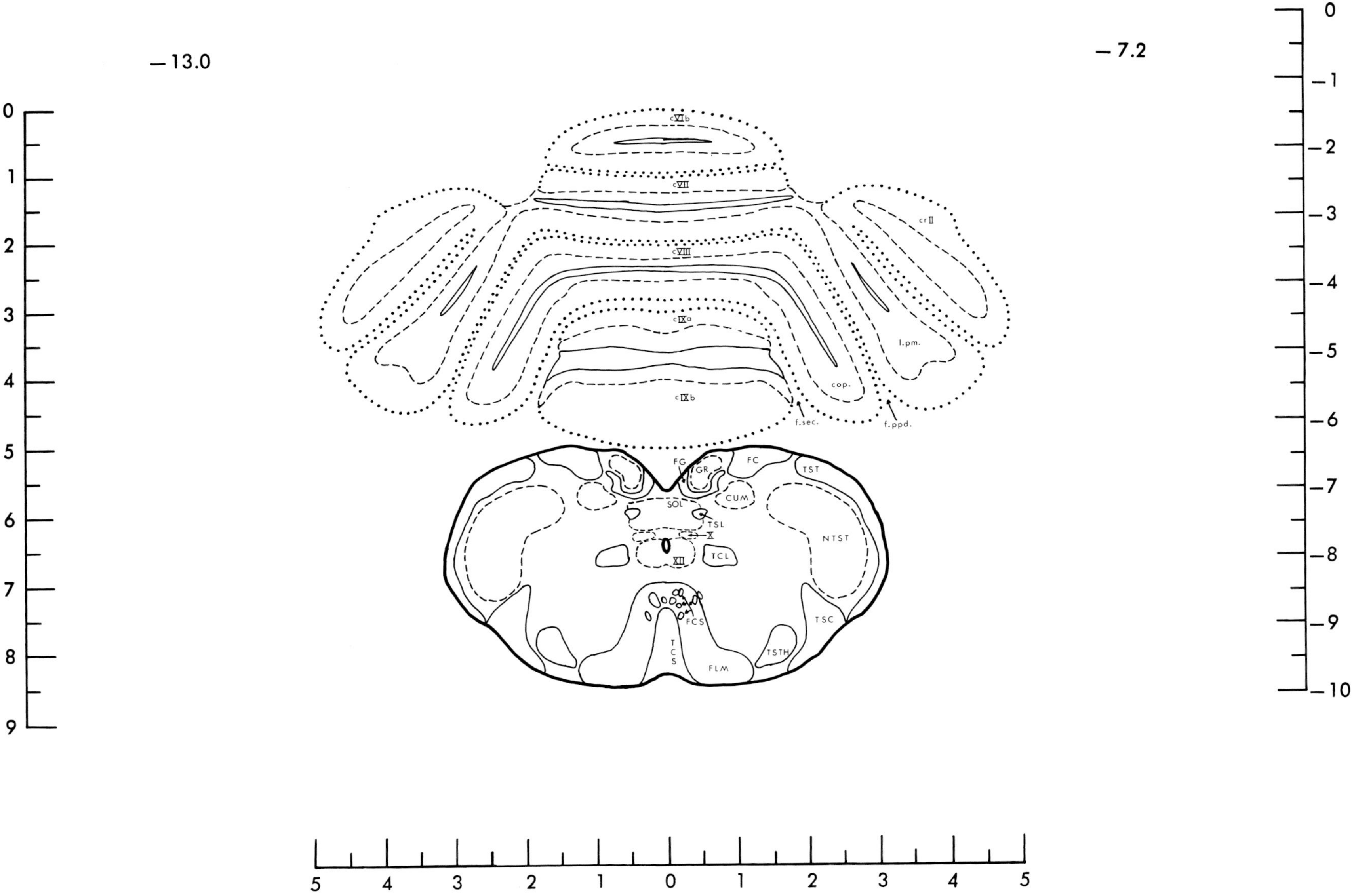

−13.0

−7.2

cVIb

cVII

cr II

cVIII

l.pm.

cIXa

cop.

cIXb

f.sec.   f.ppd.

FG GR FC
        TST
SOL CUM
    TSL
     X
XII TCL   NTST

FCS

TSC

TCS FLM TSTH

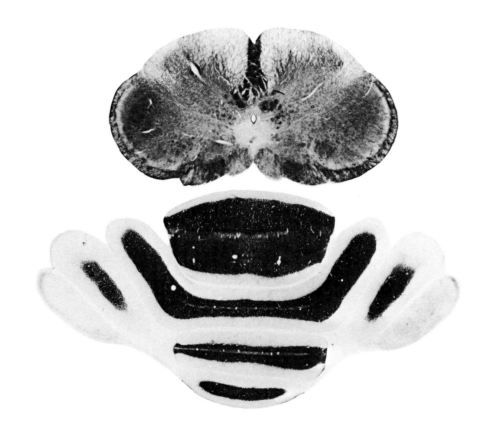

86

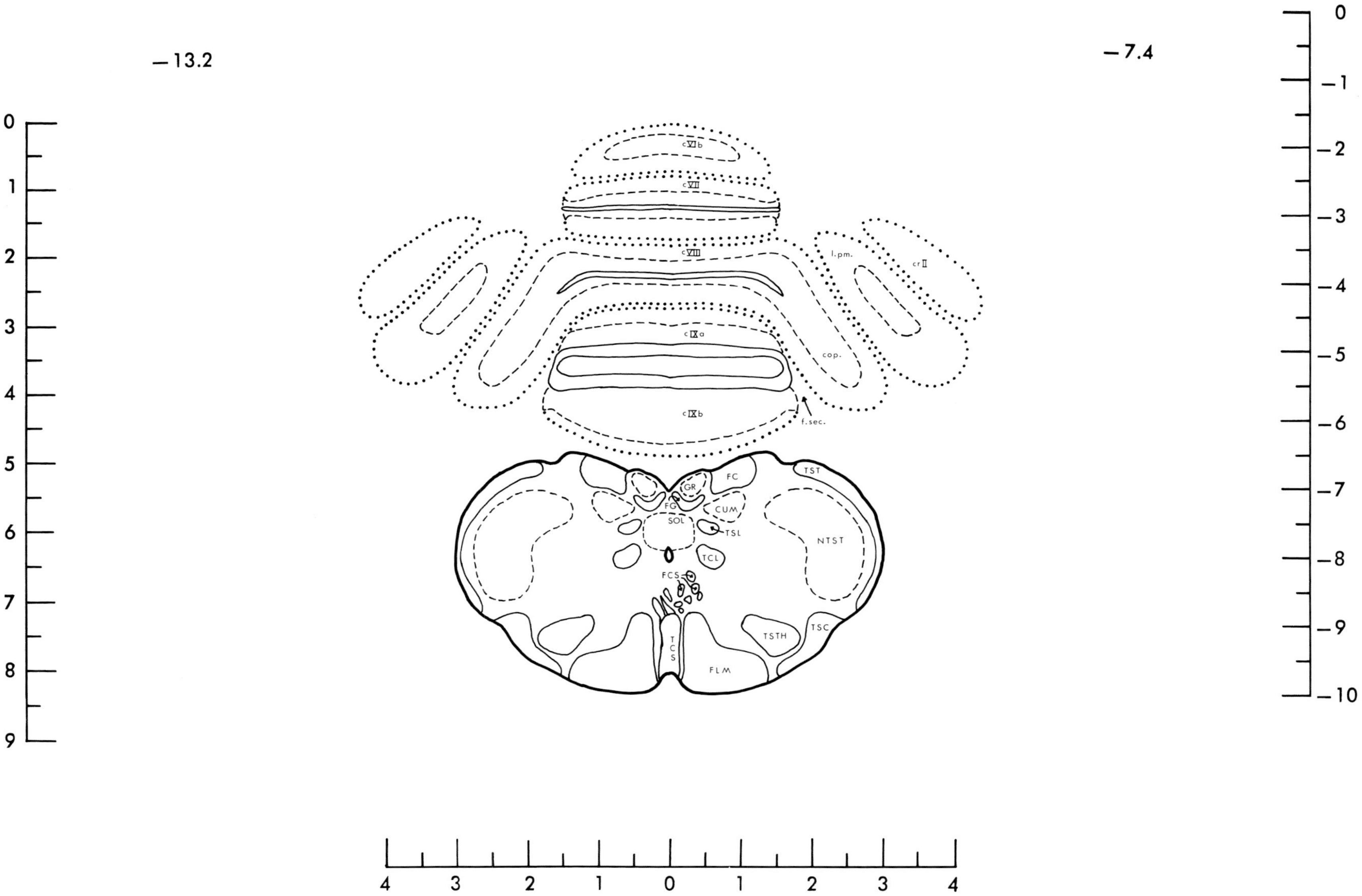

−13.2

−7.4

0
−1
−2
−3
−4
−5
−6
−7
−8
−9
−10

0
1
2
3
4
5
6
7
8
9

cⅥb
cⅦ
cⅧ
l.pm.
crⅡ
cⅨa
cop.
cⅨb
f.sec.

GR
FC
FG
CUM
TST
SOL
TSL
NTST
TCL
FCS
TCS
TSTH
TSC
FLM

4   3   2   1   0   1   2   3   4

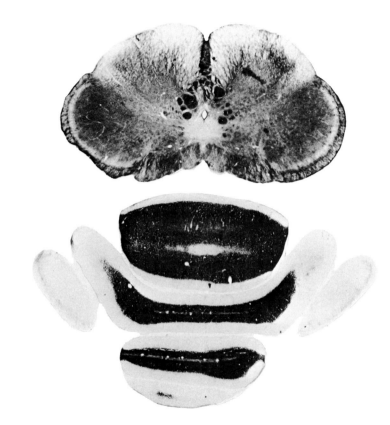

−13.4

−7.6

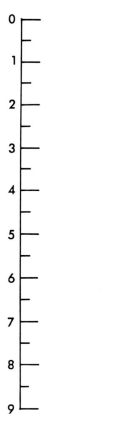

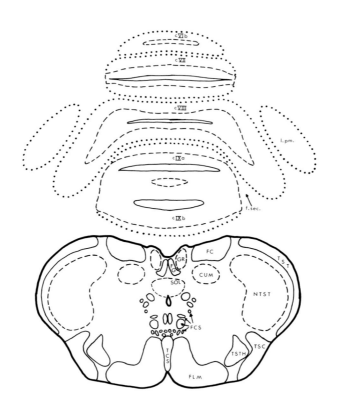

cⅦb

cⅦ

cⅧ

l.pm.

cⅨa

f.sec.

cⅨb

GR

FC

F

TST

CUM

SOL

NTST

FCS

TSC

TSTH

TCS

FLM

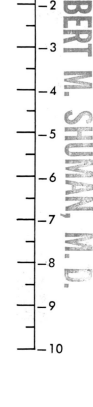

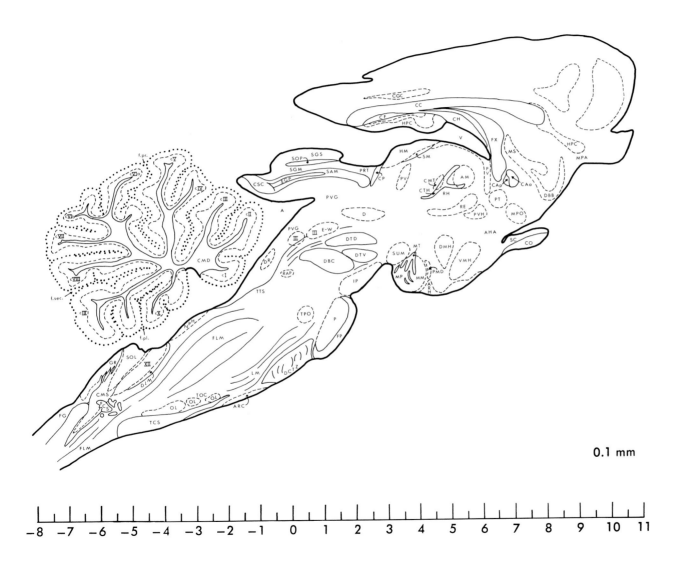

0.1 mm

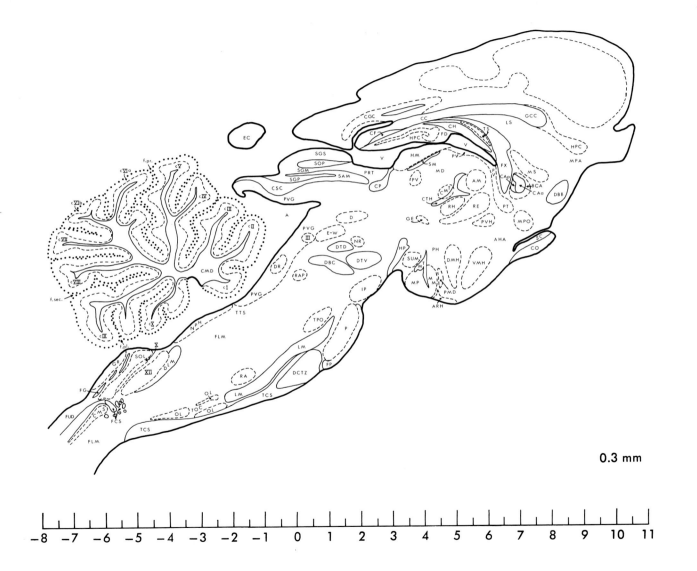

0.3 mm

102

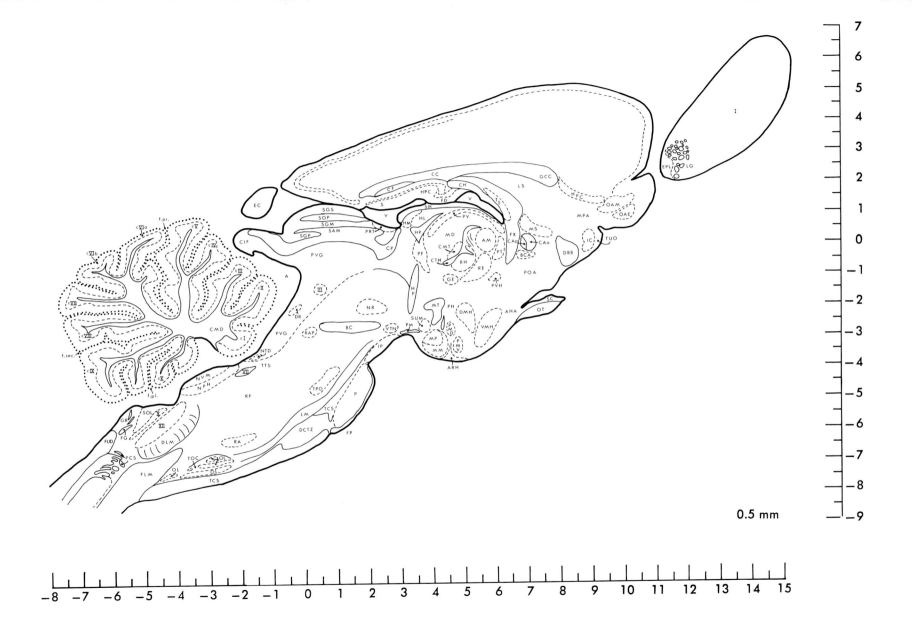

0.5 mm

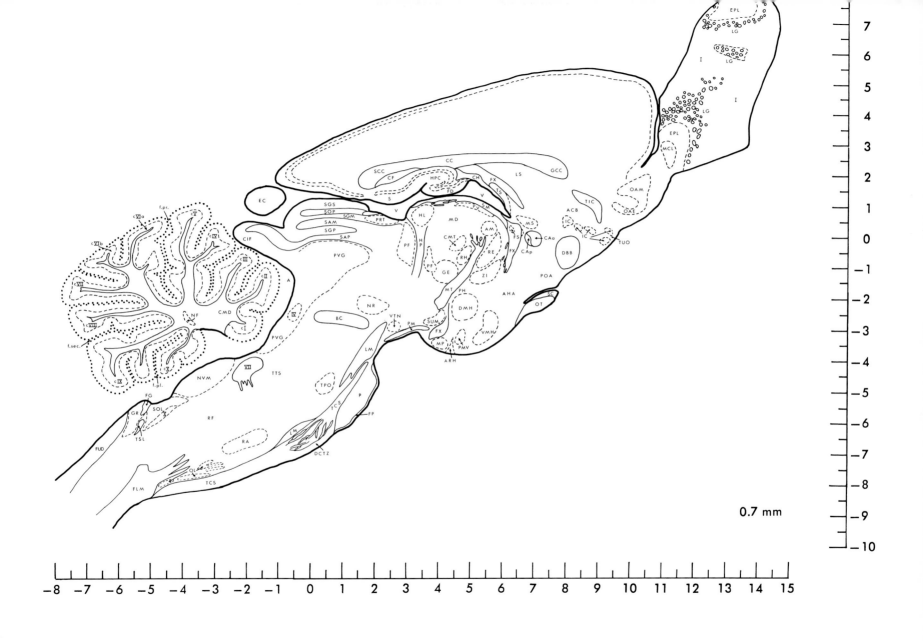

0.7 mm

104

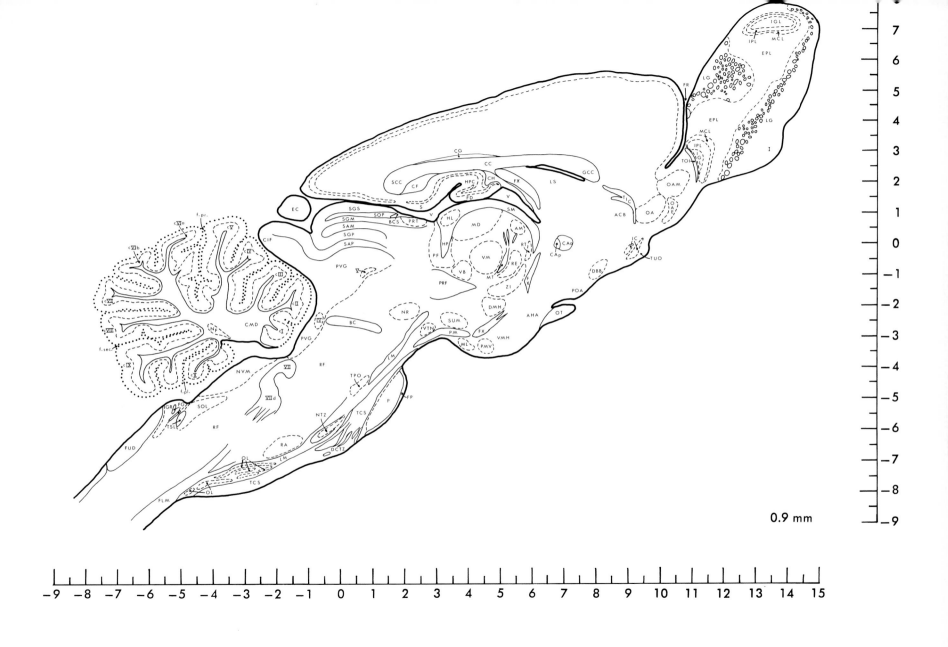

0.9 mm

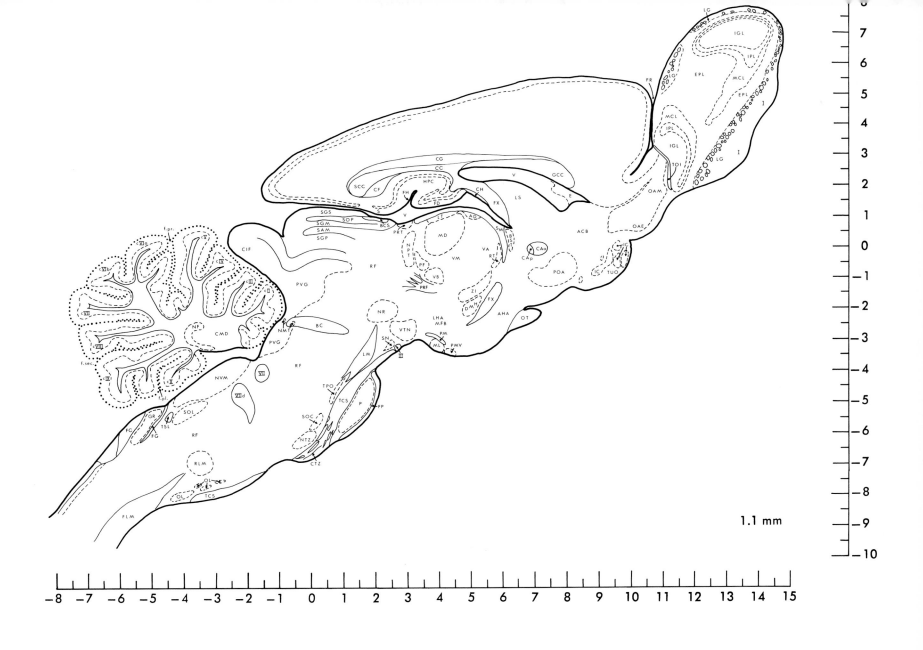

106

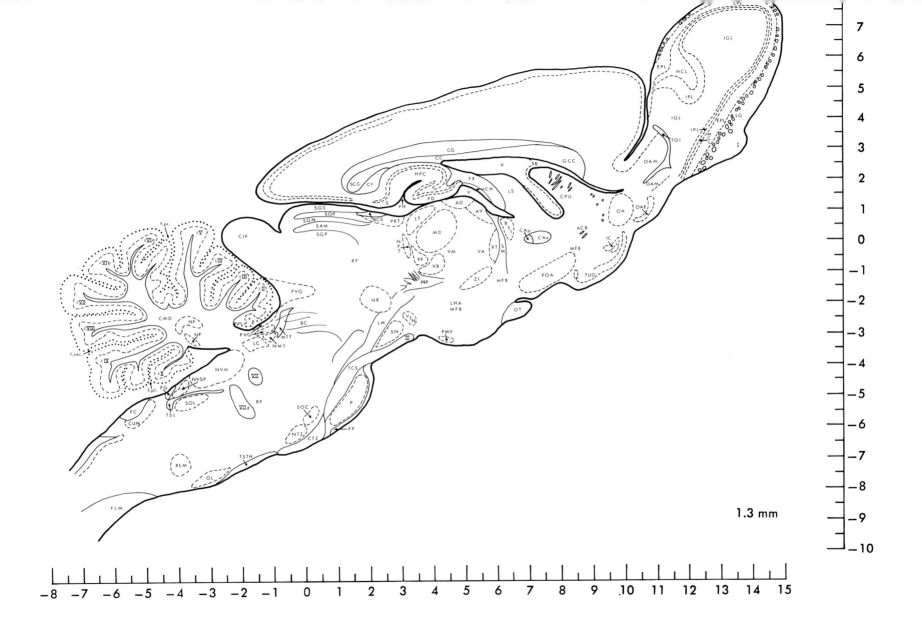

1.3 mm

107

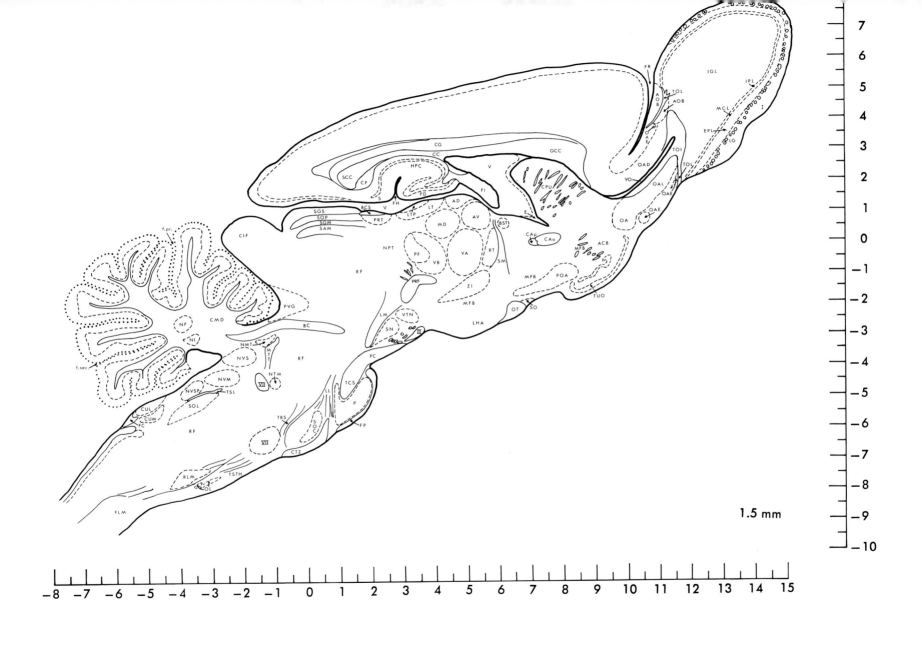

108

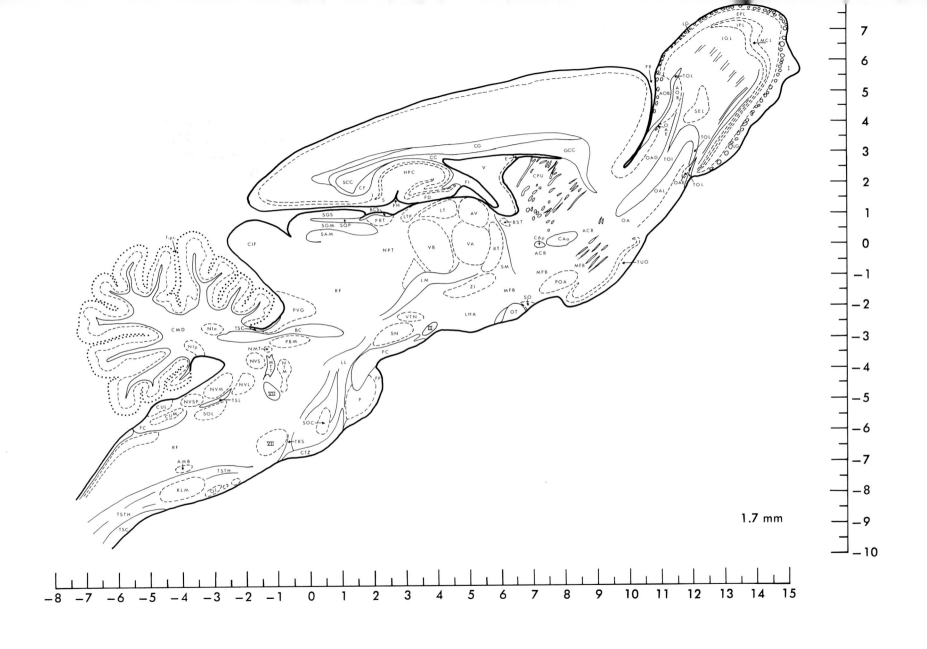

1.7 mm

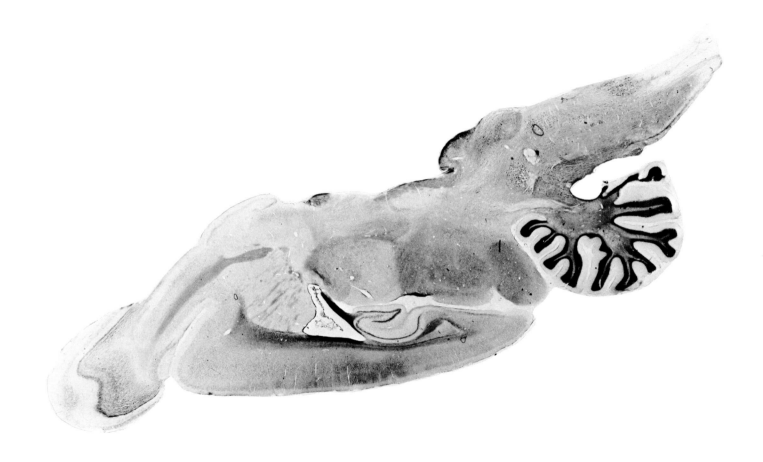

601

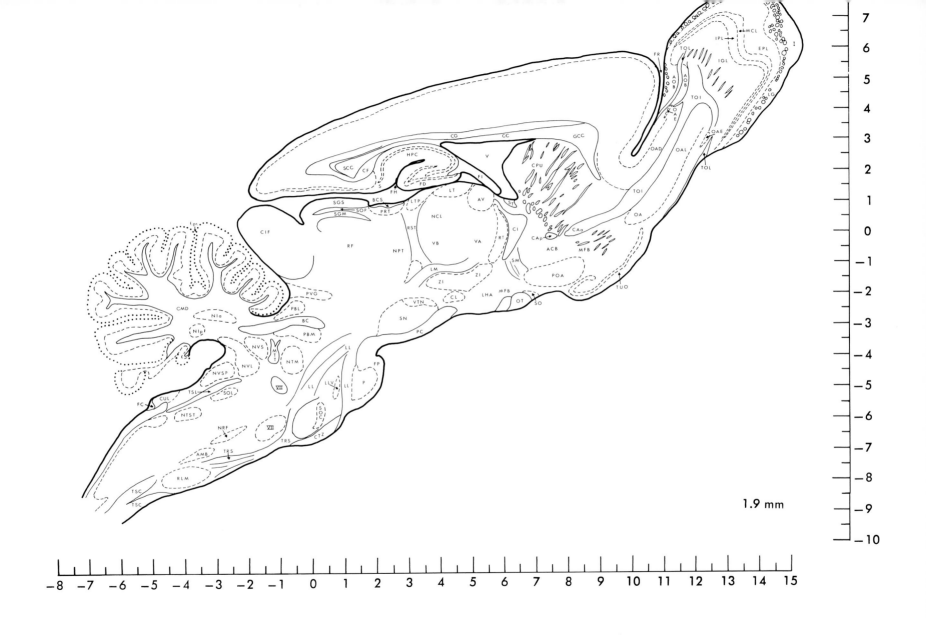

1.9 mm

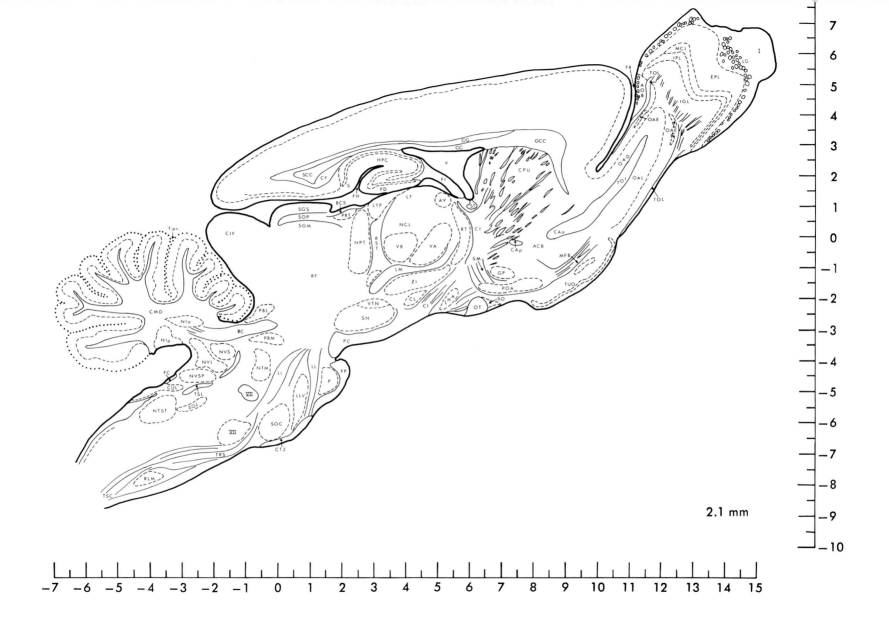

2.1 mm

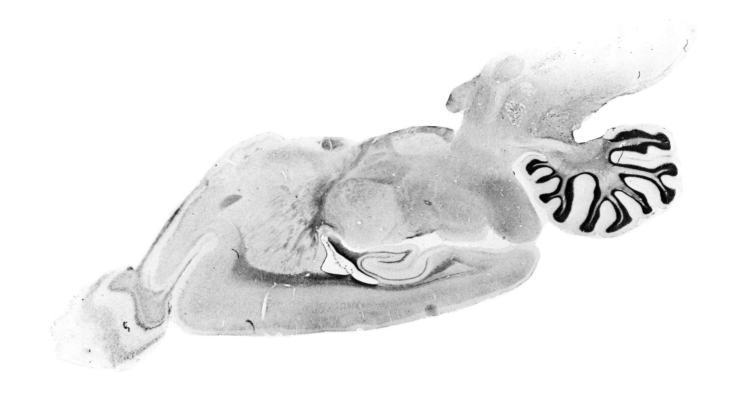

777

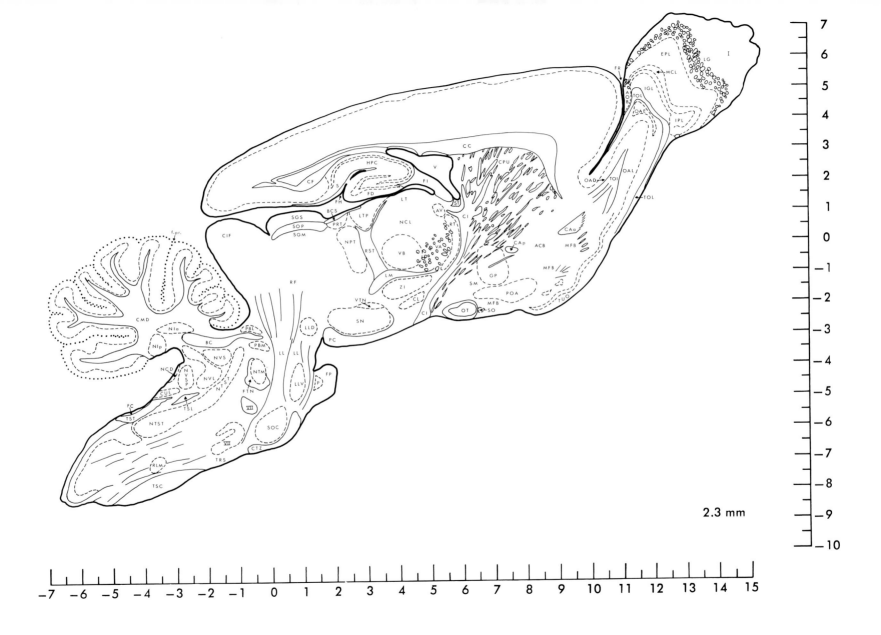

2.3 mm

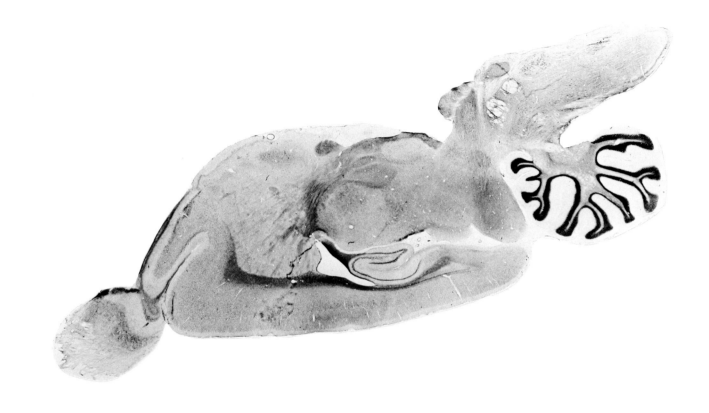

112

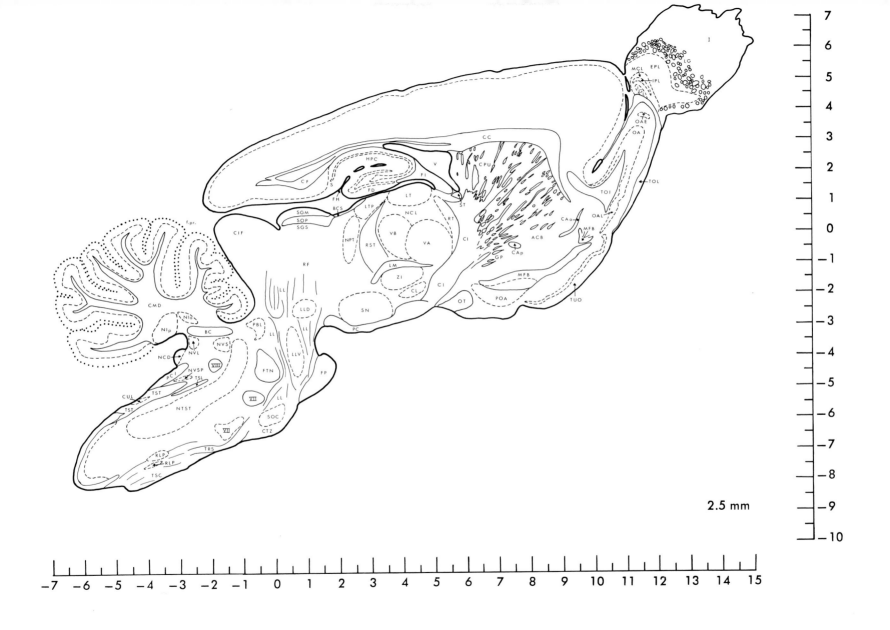

2.5 mm

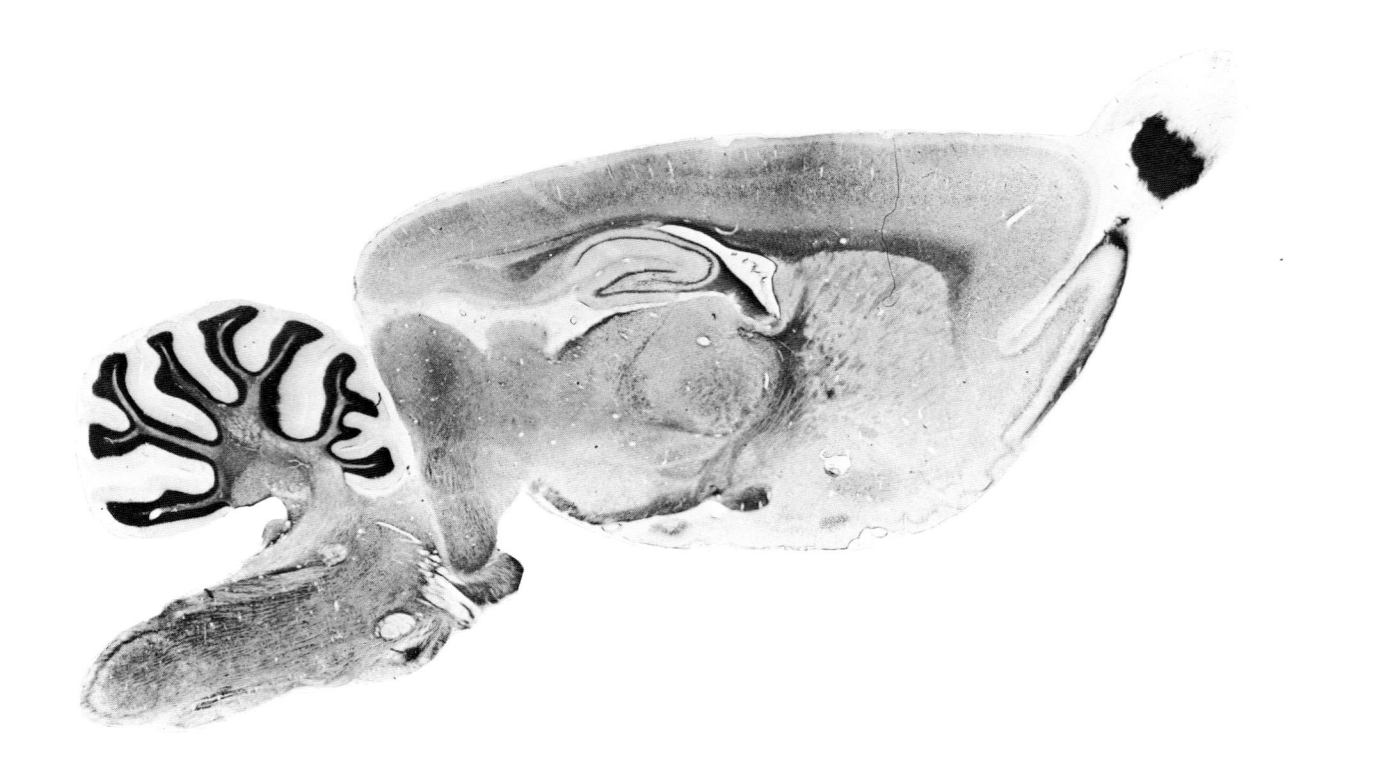

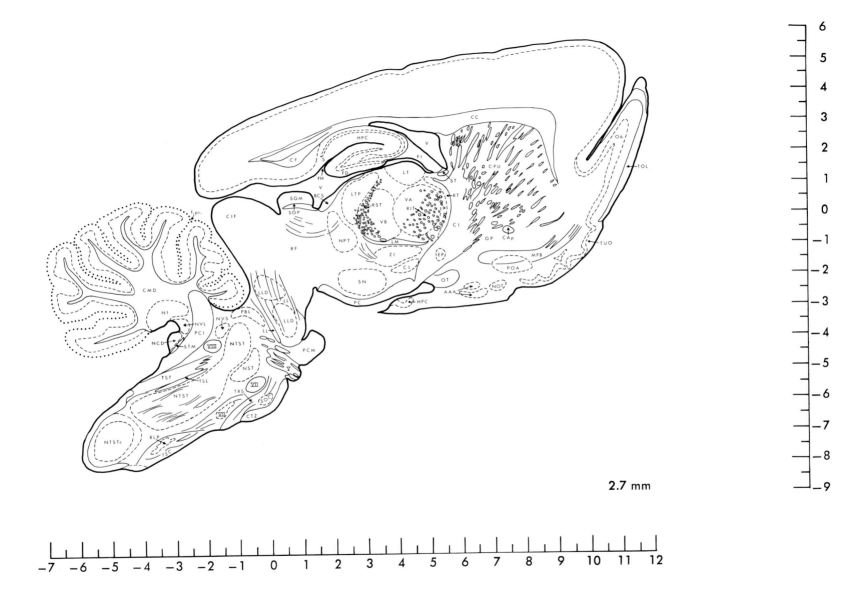

2.7 mm

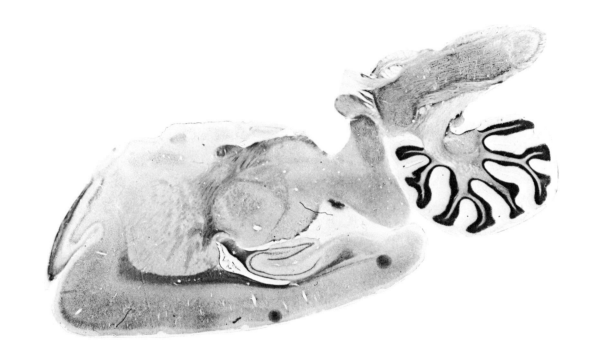

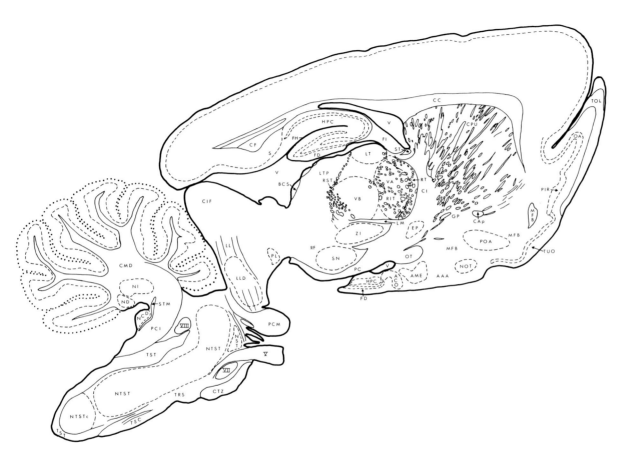

2.9 mm

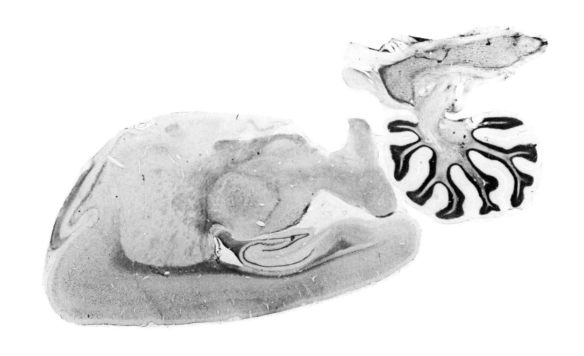

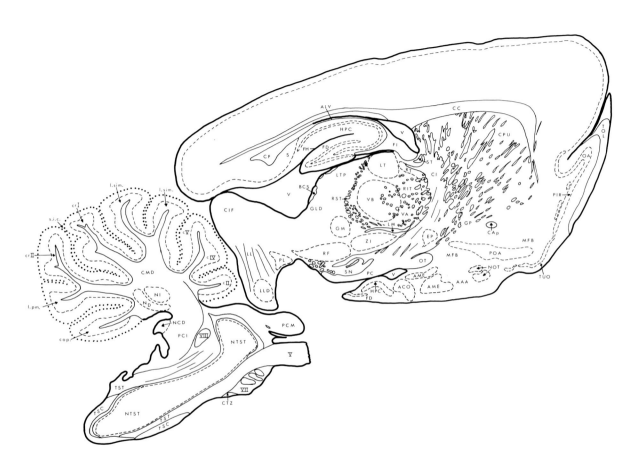

3.1 mm

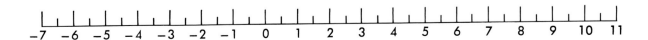

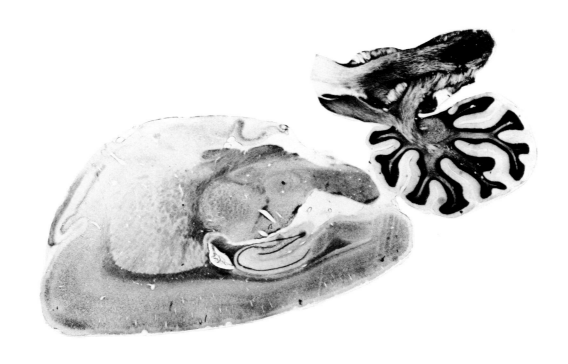

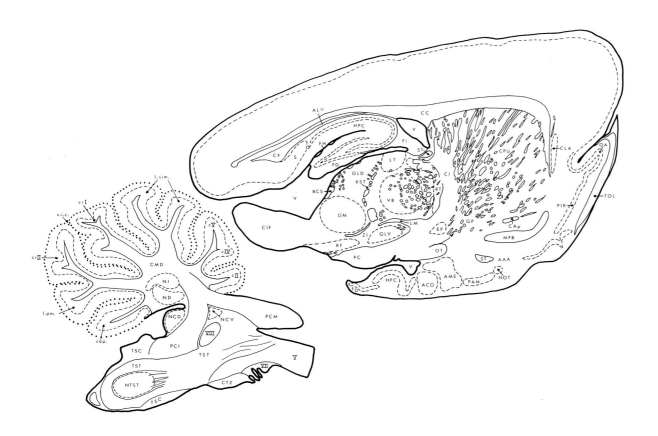

3.3 mm

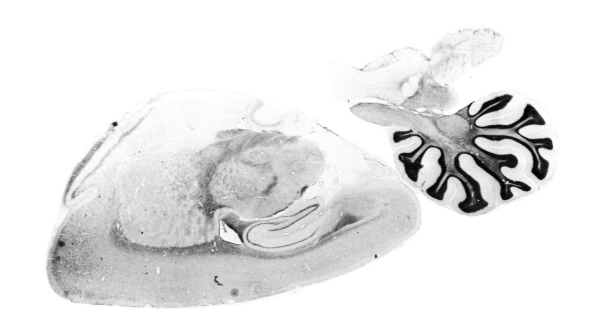

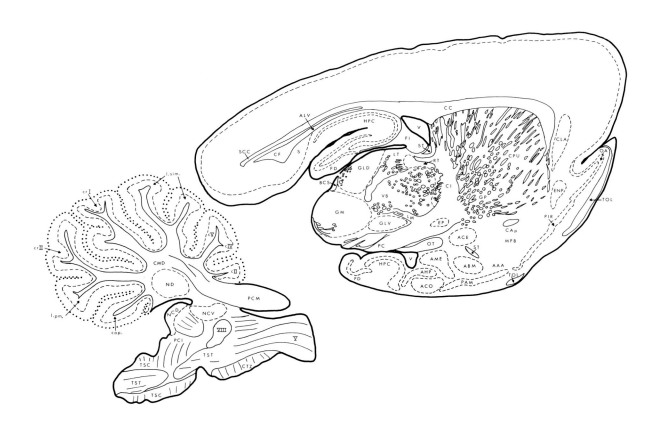

3.5 mm

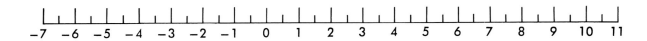

118

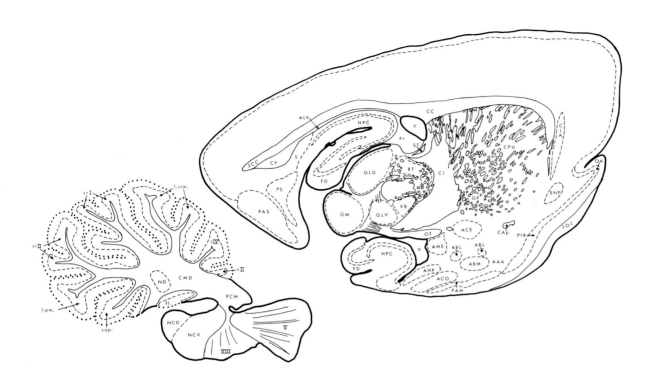

3.9 mm

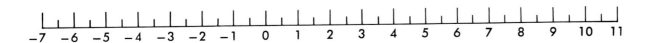

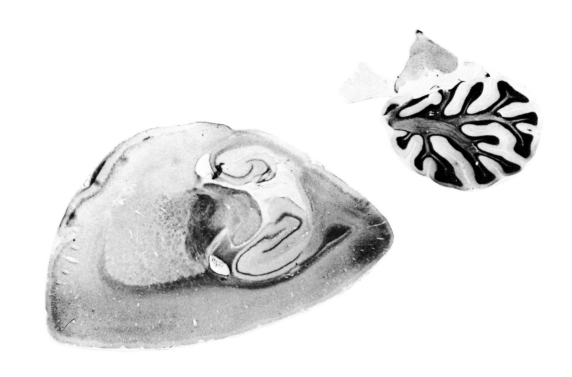

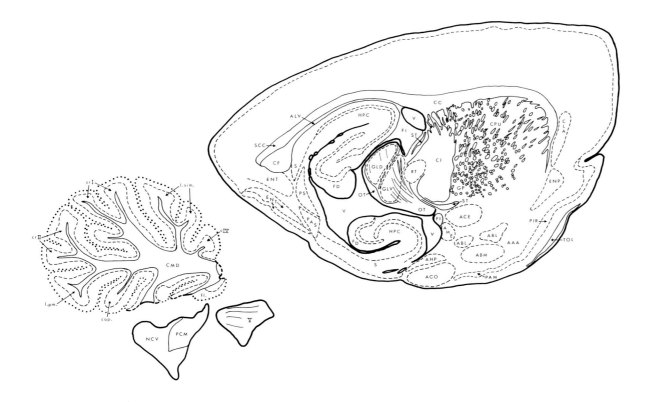
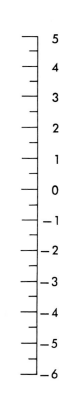

4.3 mm

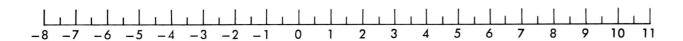

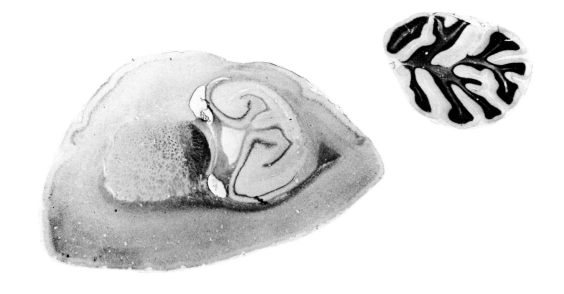

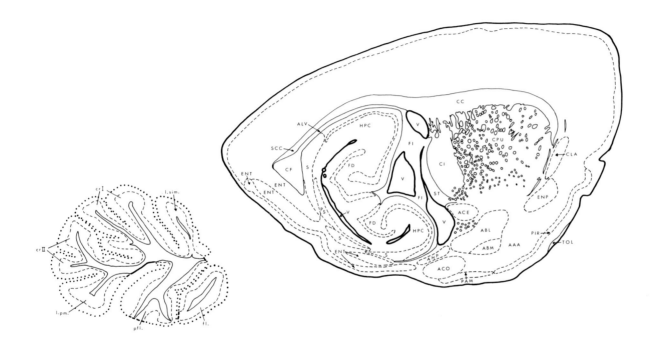

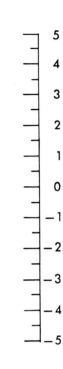

4.7 mm

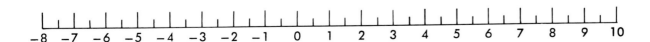

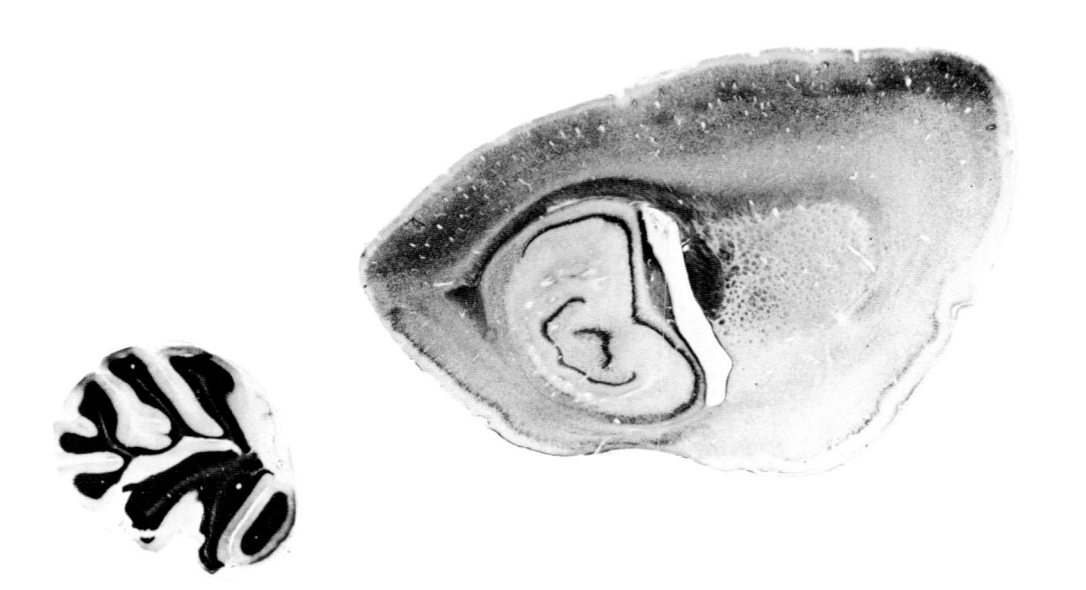

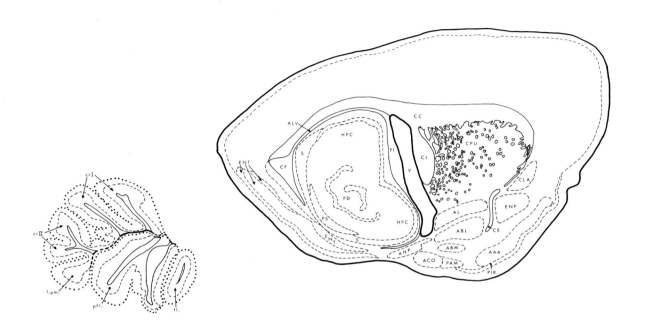

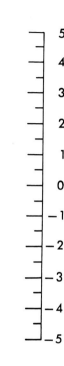

5.1 mm

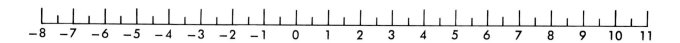

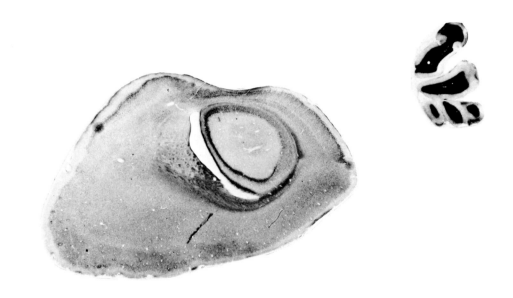

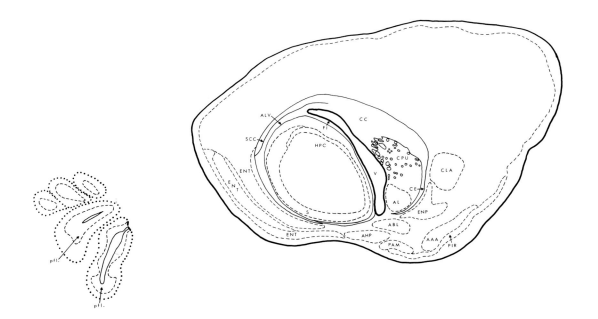
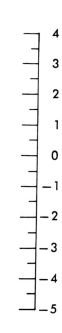

5.7 mm